Highway and Rail Transit Tunnel Maintenance and Rehabilitation Manual

2004 Edition

U.S.Department of Transportation

Federal Highway Administration
Federal Transit Administration

Notice:

The Federal Highway Administration provides high-quality information to serve Government, industry, and the public in a manner that promotes public understanding. Standards and policies are used to ensure and maximize the quality, objectivity, utility, and integrity of its information. FHWA periodically reviews quality issues and adjusts its programs and processes to ensure continuous quality improvement.

TABLE OF CONTENTS

List of Tables

List of Figures

Executive Summary

CHAPTER 1: INTRODUCTION..1-1

CHAPTER 2: TUNNEL CONSTRUCTION AND SYSTEMS 2-1

    A. Tunnel Types...2-1
        1. Shapes *2-1*
        2. Liner Types...*2-7*
        3. Invert Types...*2-8*
        4. Construction Methods *2-11*
        5. Tunnel Finishes *2-12*
    B. Ventilation Systems 2-15
        1. Types *2-15*
        2. Equipment *2-19*
    C. Lighting Systems...2-21
        1. Types *2-21*
    D. Other Systems/Appurtenances 2-22
        1. Track *2-22*
        2. Power (Third Rail/Catenary) *2-23*
        3. Signal/Communication Systems..................................*2-25*

CHAPTER 3: PREVENTIVE MAINTENANCE..3-1

    A.   Preventive Maintenance of the Tunnel Structure....................3-1
        1. Tunnel Washing..*3-1*
        2. Drain Flushing...*3-1*
        3. Ice/Snow Removal..*3-2*
        4. Tile Removal..*3-2*
    B.   Preventive Maintenance of Mechanical Systems 3-2
    C.   Preventive Maintenance of Electrical Elements 3-8
    D.   Preventive Maintenance of Track Systems 3-15
        1. Track and Supporting Structure..................................*3-15*
        2. Power (Third Rail/Catenary) *3-17*
        3. Signal/Communication Systems..................................*3-18*

E.   Preventive Maintenance of Miscellaneous Appurtenances 3-18
     1. Corrosion Protection Systems *3-18*
     2. Safety Walks, Rails, and Exit Stair/Ladder Structures *3-20*
     3. Vent Structures and Emergency Egress Shafts *3-21*

CHAPTER 4: REHABILITATION OF STRUCTURAL ELEMENTS 4-1

     A. Water Infiltration 4-1
     1. Problem *4-1*
     2. Consequences of Water Infiltration *4-2*
     3. Remediation Methods *4-3*
     B. Concrete Repairs 4-19
     1. Crack *4-20*
     2. Spall *4-23*
     C. Liner Repairs 4-29
     1. Cast-in-Place (CIP) Concrete *4-29*
     2. Pre-cast Concrete *4-30*
     3. Steel *4-30*
     4. Cast Iron *4-32*
     5. Shotcrete *4-35*
     6. Masonry *4-35*
     7. Exposed Rock *4-36*

Appendix A: Life Cycle Cost Methodology **A-1**
Glossary **G-1**
References **R-1**

LIST OF TABLES

Table 2.1 – Construction Methods... 2-11

Table 3.1 – Preventive Maintenance of Mechanical Systems 3-4

Table 3.2 – Preventive Maintenance of Electrical Systems.. 3-9

Table 4.1 – Weldability of Steel.. 4-31

LIST OF FIGURES

Figure 2.1 – Circular Highway Tunnel Shape 2-2

Figure 2.2 – Double Box Highway Tunnel Shape 2-2

Figure 2.3 – Horseshoe Highway Tunnel Shape 2-3

Figure 2.4 – Oval/Egg Highway Tunnel Shape 2-3

Figure 2.5 – Circular Rail Transit Tunnel Shape 2-4

Figure 2.6 – Double Box Rail Transit Tunnel Shape 2-5

Figure 2.7 – Single Box Rail Transit Tunnel Shape 2-5

Figure 2.8 – Horseshoe Rail Transit Tunnel Shape 2-6

Figure 2.9 – Oval Rail Transit Tunnel Shape 2-6

Figure 2.10 – Circular Tunnel Invert Type 2-9

Figure 2.11 – Single Box Tunnel Invert Type 2-10

Figure 2.12 – Horseshoe Tunnel Invert Type 2-10

Figure 2.13 – Natural Ventilation 2-15

Figure 2.14 – Longitudinal Ventilation 2-16

Figure 2.15 – Semi-Transverse Ventilation 2-17

Figure 2.16 – Full-Transverse Ventilation 2-18

Figure 2.17 – Axial Fans 2-19

Figure 2.18 – Centrifugal Fan 2-20

Figure 2.19 – Typical Third Rail Power System 2-24

Figure 2.20 – Typical Third Rail Insulated Anchor Arm 2-24

Figure 4.1 – Ice formation at location of water infiltration in plenum area above
the roadway slab 4-3

Figure 4.2 – Temporary drainage systems comprised of neoprene rubber troughs and 25 mm (1 in) aluminum channels...4-4

Figure 4.3 – Temporary drainage system comprised of 50 mm (2 in) plastic pipe......4-5

Figure 4.4 – Insulated panels used as a waterproofing lining to keep infiltrated water from freezing...4-6

Figure 4.5 – Section of membrane waterproofing system......................................4-7

Figure 4.6 – Leaking crack repair detail..4-10

Figure 4.7 – Repair of a concrete joint or crack by inclusion of a neoprene strip 4-14

Figure 4.8 – Treatment of cracks by membrane covering.......................................4-15

Figure 4.9 – Method of repairing a leaking joint 4-16

Figure 4.10 – Laser controlled cutter for removing portions of existing tunnel liner..4-18

Figure 4.11 – Horizontal surface crack repair detail...4-21

Figure 4.12 – Vertical/over head crack repair detail 4-22

Figure 4.13 – Shallow spall repair detail (shallow spall with no reinforcement steel exposed)...4-24

Figure 4.14 – Shallow spall repair detail (shallow spall with reinforcement steel exposed)...4-25

Figure 4.15 – Deep spall with exposed adequate reinforcement steel 4-27

Figure 4.16 – Deep spall with exposed inadequate reinforcement steel 4-28

Figure 4.17 – Metal Stitching Detail...4-33

Figure 4.18 – Metal Stitching Procedure..4-33

Figure 4.19 – Metal Stitching Completed 4-34

Figure 4.20 – Rock bolt types 4-37

EXECUTIVE SUMMARY

In March of 2001, the Federal Transit Administration (FTA) engaged Gannett Fleming, Inc., to develop the first ever Tunnel Management System to benefit both highway and rail transit tunnel owners throughout the United States and Puerto Rico. Specifically, these federal agencies, acting as ONE DOT, set a common goal to provide uniformity and consistency in assessing the physical condition of the various tunnel components. It is commonly understood that numerous tunnels in the United States are more than 50 years old and are beginning to show signs of considerable deterioration, especially due to water infiltration. In addition, it is desired that good maintenance and rehabilitation practices be presented that would aid tunnel owners in the repair of identified deficiencies. To accomplish these ONE DOT goals, Gannett Fleming, Inc., was tasked to produce an Inspection Manual, a Maintenance and Rehabilitation Manual, and a computerized database wherein all inventory, inspection, and repair data could be collected and stored for historical purposes.

This manual is an update to the version issued in May, 2003. It provides specific information for the maintenance and rehabilitation of both highway and rail transit tunnels. Although several components are similar in both types of tunnels, a few elements are specific to either highway or rail transits tunnels, and are defined accordingly. The following paragraphs explain the specific subjects covered along with procedural recommendations that are contained in this manual.

Introduction

This chapter presents a brief history of the project development and outlines the scope and contents of the Maintenance and Rehabilitation Manual.

Tunnel Construction and Systems

To develop uniformity concerning certain tunnel components and systems, this chapter was developed to define those major systems and describe how they relate to both highway and rail transit tunnels. This chapter is broken down into four sub-chapters that include: tunnel types, ventilation systems, lighting systems, and other systems/appurtenances.

The tunnel types section covers the different tunnel shapes in existence, liner types that have been used, the two main invert types, the various construction methods utilized to construct a tunnel, and the multiple different finishes that can be applied, mainly in highway tunnels. The ventilation and lighting system sections are self explanatory in that they cover the basic system types and configurations. The other systems/appurtenances section is used to explain tunnel systems that are present in rail transit tunnels, such as: track systems, power systems (third rail/ catenary), and signal/communications systems.

Preventive Maintenance

This chapter provides specific recommendations for performing preventive maintenance to the tunnel structure, mechanical systems, electrical elements, track systems, and miscellaneous appurtenances. The tunnel structure recommendations deal with tunnel washing, drain flushing, ice/snow removal and tile removal. The procedures for the mechanical and electrical systems/ elements are given in tabular format and include a suggested frequency for each of the tasks listed. Track systems are divided into track and supporting structure, power (third rail/catenary), and signal/communication systems. The last section for miscellaneous appurtenances covers the following three categories: 1) corrosion protection systems, 2) safety walks, rails, and exit stair/ ladder structures, and 3) vent structures and emergency egress shafts.

Rehabilitation of Structural Elements

The last chapter of this manual offers general procedural recommendations for making structural repairs to various types of tunnel liner materials. A large section is devoted to covering repairs necessary to slow, stop, or adequately divert water infiltration. Following that section is a detailed section that addresses the various structural repairs that can be made to concrete, such as repairing cracks and spalls. The last section deals with each of the following liner types: cast-in-place concrete, pre-cast concrete, steel, cast iron, shotcrete, masonry, and exposed rock.

Life-Cycle Cost Methodology

Appendix A of this manual includes a general discussion of life-cycle-cost methodology. This process could be used when determining which method of repair is most cost effective over the long term. Also, it could be used to determine if it is more beneficial to purchase a new piece of equipment or to continue maintaining the existing piece.

CHAPTER 1:
INTRODUCTION

Background

In 1999, the Federal Highway Administration (FHWA) created an office to focus on management of highway assets. Part of this office is responsible for providing guidance and technical assistance to state and local highway agencies on structure management issues, including highway tunnels. Similarly, the Federal Transit Administration (FTA) is responsible for providing guidance on tunnel management to rail transit owners. Because of this common interest in tunnel management procedures, the two agencies decided to jointly sponsor the development of a Tunnel Management System for both highway and rail transit tunnel owners.

To avoid future potential major operation problems due to deferred maintenance, FHWA and FTA sponsored this project to develop inspection procedures and guidance for maintenance practices within highway and rail transit tunnels and to assist tunnel owners in maintaining their tunnels. Along with the Inspection Manual and this companion Maintenance and Rehabilitation Manual, a computerized database system was also developed to assist with the storage and management of tunnel condition data and for prioritizing repairs. It is the intent of the FHWA and FTA that these products be furnished to each highway and rail transit tunnel owner across the nation, and to be placed in the public domain.

Phase 1 of this project involved the development of an inventory database of the nation's highway and rail transit tunnels that included such information as location of the tunnel, tunnel name, age, length, shape, height, width, the construction method employed, construction ground conditions, lining/support types, and types of mechanical/electrical systems. The data received from highway tunnel owners responding to the questionnaire revealed that more than 32 percent of reported highway tunnels are between 50-100 years old, with 4 percent greater than 100 years old. Although it is more difficult to categorize rail transit tunnels by percent, inventory information collected to date, plus data known to exist for certain agencies that had trouble segmenting all of their tunnels according to the questionnaire, suggests that there are approximately 346 km (215 miles) of rail transit tunnels greater than 50 years old. This data is sufficient to indicate that these older highway and rail transit tunnels contain elements that are deteriorating and in need of repair.

Groundwater infiltration through joints and cracks in tunnels is the number one cause of deterioration of the various tunnel elements. In addition, for concrete tunnels more than 50 years in age it is highly likely that the concrete was not air-entrained and; therefore, tunnels subjected to temperature gradients may have suffered damage over the years due to freeze-thaw actions. Since numerous tunnels have been subjected to these conditions for many years, it is vitally important that tunnel owners commence regular preventive maintenance and repair procedures for correcting deficiencies such that each tunnel can continue to function as originally designed.

Scope

The purpose of this manual is to provide highway and rail transit tunnel owners with guidelines and practices for preventive maintenance of both the tunnel structure and the mechanical/electrical/track systems within. Suggested repairs to the tunnel structure for various deficiencies are provided. These repairs include guidelines for controlling water infiltration into the tunnel, the number one cause of deterioration.

Contents

To promote consistency of definition of particular elements, this manual contains several chapters that explain the various types of elements that exist within the tunnel. For example, the description of tunnel components such as tunnel configuration, liner types, invert types, ventilation systems, lighting systems, tunnel finishes and other systems/appurtenances (track, traction power, signals and communications) are each provided in separate sections to assist tunnel owners in educating their inspectors as to the particular system existing within the tunnel.

The incorporation of the guidelines presented herein and the use of a documented maintenance and inspection program (via the software provided) will help tunnel owners to program needed maintenance and rehabilitation costs. It is important to note that the guidelines and practices included are intended to supplement existing programs and procedures already in place. It is not the intent to replace current practices unless the tunnel owner decides to do so as a benefit to his/her program.

CHAPTER 2:
TUNNEL CONSTRUCTION AND SYSTEMS

A.   TUNNEL TYPES

This section describes the various types of highway and rail transit tunnels. These tunnel types are described by their shape, liner type, invert type, construction method, and tunnel finishes. It should be noted that other types may exist currently or be constructed in the future as new technologies become available. The purpose of this section is to look at the types that are most commonly used in tunnel construction to help the inspector properly classify any given tunnel. As a general guideline a minimum length of 100 meters (~300 feet) was used in defining a tunnel for inventory purposes. This length is primarily to exclude long underpasses, however other reasons for using the tunnel classification may exist such as the presence of lighting or a ventilation system, which could override the length limitation.

1.   Shapes

a)   Highway Tunnels

As shown in Figures 2.1 to 2.4, there are four main shapes of highway tunnels – circular, rectangular, horseshoe, and oval/egg. The different shapes typically relate to the method of construction and the ground conditions in which they were constructed. Although many tunnels will appear rectangular from inside, due to horizontal roadways and ceiling slabs, the outside shape of the tunnel defines its type. Some tunnels may be constructed using combinations of these types due to different soil conditions along the length of the tunnel. Another possible highway tunnel shape that is not shown is a single box with bi-directional traffic.

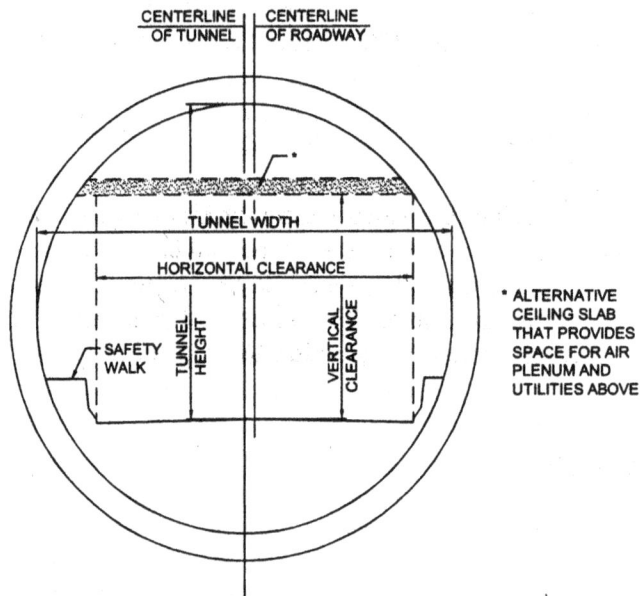

Figure 2.1 – Circular tunnel with two traffic lanes and one safety walk. Also shown is an alternative ceiling slab. Invert may be solid concrete over liner or a structural slab.

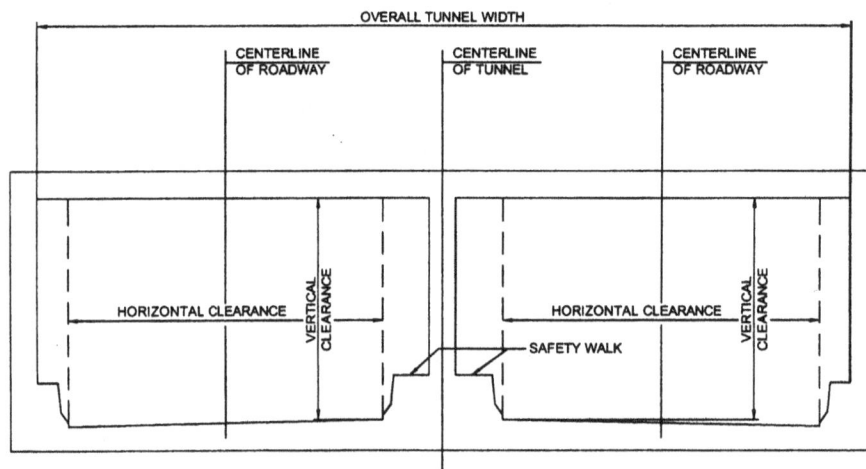

Figure 2.2 – Double box tunnel with two traffic lanes and one safety walk in each box. Depending on location and loading conditions, the center wall may be solid or composed of consecutive columns.

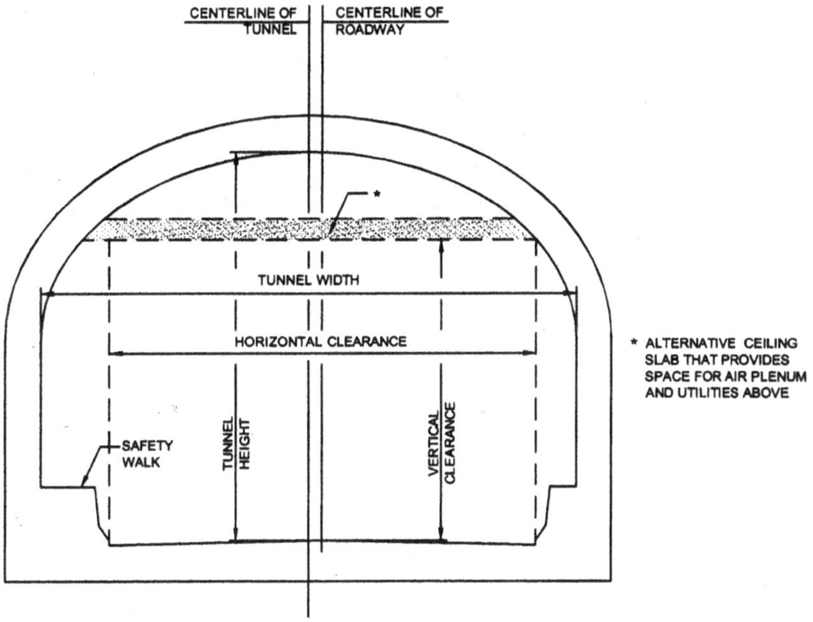

Figure 2.3 – Horseshoe tunnel with two traffic lanes and one safety walk. Also shown is an alternative ceiling slab. Invert may be a slab on grade or a structural slab.

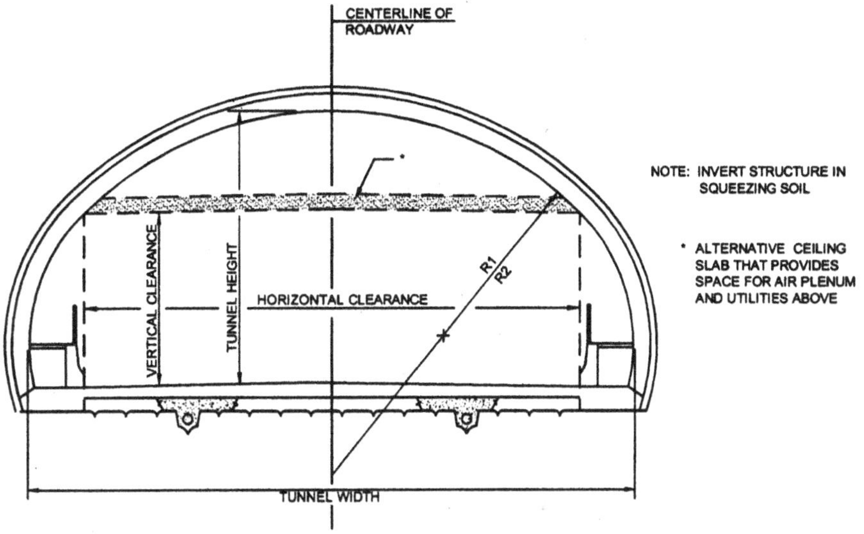

Figure 2.4 – Oval/egg tunnel with three traffic lanes and two safety walks. Also shown is an alternative ceiling slab.

b)   Rail Transit Tunnels

Figures 2.5 to 2.9 show the typical shapes for rail transit tunnels. As with highway tunnels, the shape typically relates to the method/ground conditions in which they were constructed. The shape of rail transit tunnels often varies along a given rail line. These shapes typically change at the transition between the station structure and the typical tunnel cross-section. However, the change in shape may also occur between stations due to variations in ground conditions.

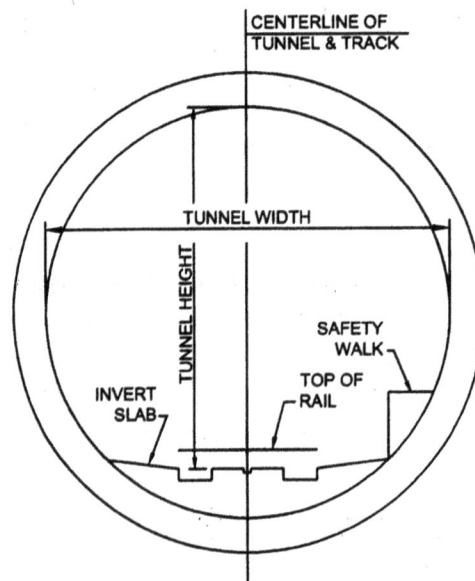

Figure 2.5 – Circular tunnel with a single track and one safety walk. Invert slab is placed on top of liner.

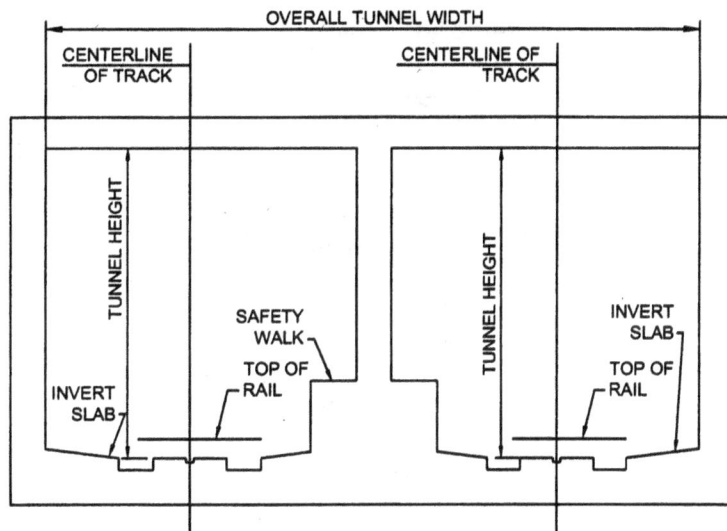

Figure 2.6 – Double box tunnel with a single track and one safety walk in each box. Depending on location and loading conditions, center wall may be solid or composed of consecutive columns.

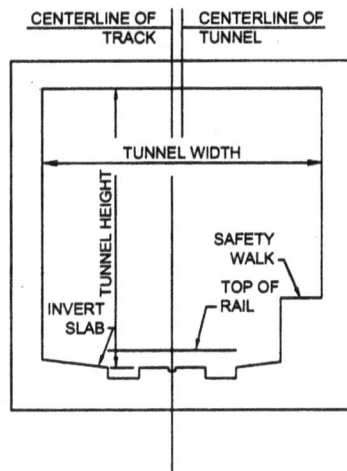

Figure 2.7 – Single box tunnel with a single track and one safety walk. Tunnel is usually constructed beside another single box tunnel for opposite direction travel.

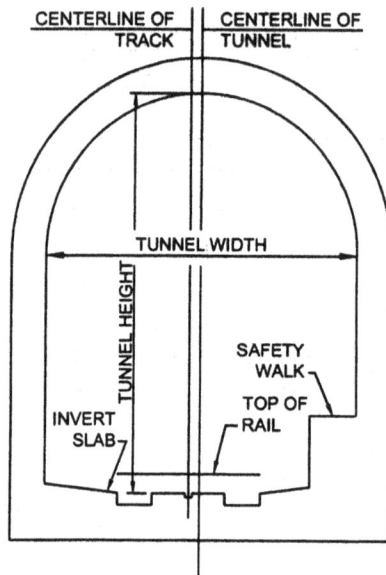

Figure 2.8 – Horseshoe tunnel with a single track and one safety walk. This shape typically exists in rock conditions and may be unlined within stable rock formations.

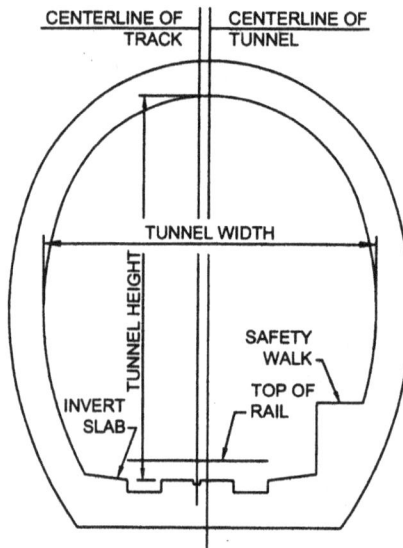

Figure 2.9 – Oval tunnel with a single track and one safety walk.

2. Liner Types

Tunnel liner types can be described using the following classifications:

- Unlined Rock
- Rock Reinforcement Systems
- Shotcrete
- Ribbed Systems
- Segmental Linings
- Placed Concrete
- Slurry Walls.

a) Unlined Rock

As the name suggests, an unlined rock tunnel is one in which no lining exists for the majority of the tunnel length. Linings of other types may exist at portals or at limited zones of weak rock. This type of liner was common in older railroad tunnels in the western mountains, some of which have been converted into highway tunnels for local access.

b) Rock Reinforcement Systems

Rock reinforcement systems are used to add additional stability to rock tunnels in which structural defects exist in the rock. The intent of these systems is to unify the rock pieces to produce a composite resistance to the outside forces. Reinforcement systems include the use of metal straps and mine ties with short bolts, untensioned steel dowels, or tensioned steel bolts. To prevent small fragments of rock from spalling off the lining, wire mesh, shotcrete, or a thin concrete lining may be used in conjunction with the above systems.

c) Shotcrete

Shotcrete is appealing as a lining type due to its ease of application and short "stand-up" time. Shotcrete is primarily used as a temporary application prior to a final liner being installed or as a local solution to instabilities in a rock tunnel. However, shotcrete can be used as a final lining. When this is the case, it is typically placed in layers and can have metal or randomly oriented, synthetic fibers as reinforcement. The inside surface can be finished smooth as with regular concrete; therefore, it is difficult to determine the lining type without having knowledge of the construction method.

d) Ribbed Systems

Ribbed systems are typically a two-pass system for lining a drill-and-blast rock tunnel. The first pass consists of timber, steel, or precast concrete ribs usually with blocking between them. This provides structural stability to the tunnel. The second pass typically consists of poured concrete that is placed inside of the ribs. Another application of this system is to form the ribs using prefabricated reinforcing bar cages embedded in multiple layers of shotcrete. One other soft ground application is to place "barrel stave" timber lagging between the ribs.

e)   Segmental Linings

Segmental linings are primarily used in conjunction with a tunnel boring machine (TBM) in soft ground conditions. The prefabricated lining segments are erected within the cylindrical tail shield of the TBM. These prefabricated segments can be made of steel, concrete, or cast iron and are usually bolted together to compress gaskets for preventing water penetration.

f)   Placed Concrete

Placed concrete linings are usually the final linings that are installed over any of the previous initial stabilization methods. They can be used as a thin cover layer over the primary liner to provide a finished surface within the tunnel or to sandwich a waterproofing membrane. They can be reinforced or unreinforced. They can be designed as a non-structural finish element or as the main structural support for the tunnel.

g)   Slurry Walls

Slurry wall construction types vary, but typically they consist of excavating a trench that matches the proposed wall profile. This trench is continually kept full with a drilling fluid during excavation, which stabilizes the sidewalls. Then a reinforcing cage is lowered into the slurry or soldier piles are driven at a predetermined interval and finally tremie concrete is placed into the excavation, which displaces the drilling fluid. This procedure is repeated in specified panel lengths, which are separated with watertight joints.

3.   **Invert Types**

The invert of a tunnel is the slab on which the roadway or track bed is supported. There are two main methods for supporting the roadway or track bed; one is by placing the roadway or track bed directly on grade at the bottom of the tunnel structure, and the other is to span the roadway between sidewalls to provide space under the roadway for ventilation and utilities. The first method is used in most rail transit tunnels because their ventilation systems rarely use supply ductwork under the slab. This method is also employed in many highway tunnels over land where ventilation is supplied from above the roadway level.

The second method is commonly found in circular highway tunnels that must provide a horizontal roadway surface that is wide enough for at least two lanes of trafficand therefore the roadway slab is suspended off the tunnel bottom a particular distance. The void is then used for a ventilation plenum and other utilities. The roadway slab in many of the older highway tunnels in New York City is supported by placing structural steel beams, encased in concrete, that span transversely to the tunnel length, and are spaced between 750 mm (30 in) and 1,500 mm (60 in) on centers. Newer tunnels, similar to the second Hampton Roads Tunnel in Virginia, provide structural reinforced concrete slabs that span the required distance between supports.

It is necessary to determine the type of roadway slab used in a given tunnel because a more extensive inspection is required for a structural slab than for a slab-on-grade. Examples of structural slabs in common tunnel shapes are shown in Figures 2.10 to 2.12.

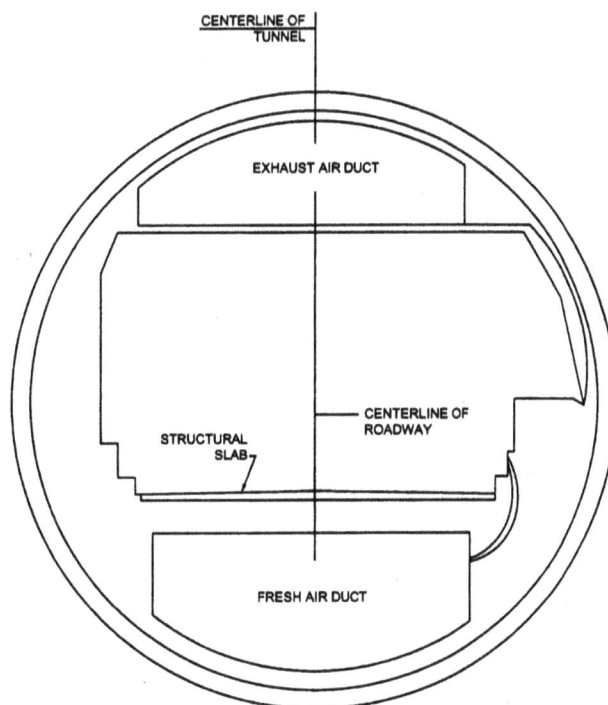

Figure 2.10 – Circular tunnel with a structural slab that provides space for an air plenum below.

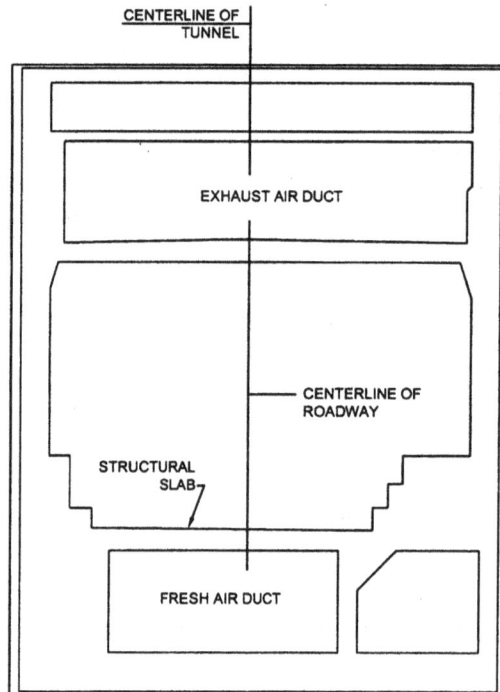

Figure 2.11 – Single box tunnel with a structural slab that provides space for an air plenum below.

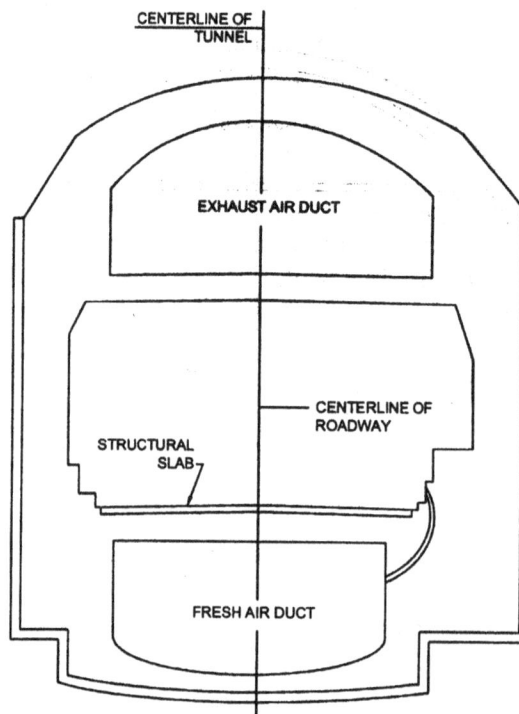

Figure 2.12 – Horseshoe tunnel with a structural slab that provides space for an air plenum below.

4. Construction Methods

As mentioned previously, the shape of the tunnel is largely dependent on the method used to construct the tunnel. Table 2.1 lists the seven main methods used for tunnel construction with the shape that typically results. Brief descriptions of the construction methods follow:

Table 2.1 – Construction Methods

	Circular	Horseshoe	Rectangular
Cut and Cover			X
Shield Driven	X		
Bored	X		
Drill and Blast	X	X	
Immersed Tube	X		X
Sequential Excavation		X	
Jacked Tunnel	X		X

a) Cut and Cover

This method involves excavating an open trench in which the tunnel is constructed to the design finish elevation and subsequently covered with various compacted earthen materials and soils. Certain variations of this method include using piles and lagging, tie back anchors or slurry wall systems to construct the walls of a cut and cover tunnel.

b) Shield Driven

This method involves pushing a shield into the soft ground ahead. The material inside the shield is removed and a lining system is constructed before the shield is advanced further.

c) Bored

This method refers to using a mechanical TBM in which the full face of the tunnel cross section is excavated at one time using a variety of cutting tools that depend on ground conditions (soft ground or rock). The TBM is designed to support the adjacent soil until temporary (and subsequently permanent) linings are installed.

d) Drill and Blast

An alternative to using a TBM in rock situations would be to manually drill and blast the rock and remove it using conventional conveyor techniques. This method was commonly used for older tunnels and is still used when it is determined cost effective or in difficult ground conditions.

e)   Immersed Tube

When a canal, channel, river, etc., needs to be crossed, this method is often used. A trench is dug at the water bottom and prefabricated tunnel segments are made water tight and sunken into position where they are connected to the other segments. Afterward, the trench may be backfilled with earth to cover and protect the tunnel from the water-borne traffic, e.g., ships, barges, and boats.

f)   Sequential Excavation Method (SEM)

Soil in certain tunnels may have sufficient strength such that excavation of the soil face by equipment in small increments is possible without direct support. This excavation method is called the sequential excavation method. Once excavated, the soil face is then supported using shotcrete and the excavation is continued for the next segment. The cohesion of the rock or soil can be increased by injecting grouts into the ground prior to excavation of that segment.

g)   Jacked Tunnels

The method of jacking a large tunnel underneath certain obstructions (highways, buildings, rail lines, etc.) that prohibit the use of typical cut-and-cover techniques for shallow tunnels has been used successfully in recent years. This method is considered when the obstruction cannot be moved or temporarily disturbed. First jacking pits are constructed. Then tunnel sections are constructed in the jacking pit and forced by large hydraulic jacks into the soft ground, which is systematically removed in front of the encroaching tunnel section. Sometimes if the soil above the proposed tunnel is poor then it is stabilized through various means such as grouting or freezing.

5.   **Tunnel Finishes**

The interior finish of a tunnel is very important to the overall tunnel function. The finishes must meet the following standards to ensure tunnel safety and ease of maintenance:

- Be designed to enhance tunnel lighting and visibility
- Be fire resistant
- Be precluded from producing toxic fumes during a fire
- Be able to attenuate noise
- Be easy to clean.

A brief description of the typical types of tunnel finishes that exist in highway tunnels is given below. Transit tunnels often do not have an interior finish because the public is not exposed to the tunnel lining except as the tunnel approaches the stations or portals.

a)   Ceramic Tile

This type of tunnel finish is the most widely used by tunnel owners. Tunnels with a concrete or shotcrete inner lining are conducive to tile placement because of their smooth surface. Ceramic tiles are extremely fire resistant, economical, easily cleaned, and good reflectors of light due to the smooth, glazed exterior finish. They are not; however, good sound attenuators, which in new tunnels has been addressed through other means. Typically, tiles are 106 mm (4 ¼ in) square and are available in a wide variety of colors. They differ from conventional ceramic tile in that they require a more secure connection to the tunnel lining to prevent the tiles from falling onto the roadway below. Even with a more secure connection, tiles may need to be replaced eventually because of normal deterioration. Additional tiles are typically purchased at the time of original construction since they are specifically made for that tunnel. The additional amount purchased can be up to 10 percent of the total tiled surface.

b)   Porcelain-Enameled Metal Panels

Porcelain enamel is a combination of glass and inorganic color oxides that are fused to metal under extremely high temperatures. This method is used to coat most home appliances. The Porcelain Enamel Institute (PEI) has established guidelines for the performance of porcelain enamel through the following publications:
- Appearance Properties (PEI 501)
- Mechanical and Physical Properties (PEI 502)
- Resistance to Corrosion (PEI 503)
- High Temperature Properties (PEI 504)
- Electrical Properties (PEI 505).

Porcelain enamel is typically applied to either cold-formed steel panels or extruded aluminum panels. For ceilings, the panels are often filled with a lightweight concrete; for walls, fiberglass boards are frequently used. The attributes of porcelain-enameled panels are similar to those for ceramic tile previously discussed; they are durable, easily washed, reflective, and come in a variety of colors. As with ceramic tile, these panels are not good for sound attenuation.

c)   Epoxy-Coated Concrete

Epoxy coatings have been used on many tunnels during construction to reduce costs. Durable paints have also been used. The epoxy is a thermosetting resin that is chemically formulated for its toughness, strong adhesion, reflective ability, and low shrinkage. Experience has shown that these coatings do not withstand the harsh tunnel environmental conditions as well as the others, resulting in the need to repair or rehabilitate more often.

d)   Miscellaneous Finishes

There are a variety of other finishes that can be used on the walls or ceilings of tunnels. Some of these finishes are becoming more popular due to their improved sound absorptive properties, ease of replacement, and ability to capitalize on the benefits of some of the materials mentioned above. Some of the systems are listed below:

(1)   Coated Cementboard Panels

These panels are not in wide use in American tunnels at this time, but they offer a lightweight, fiber-reinforced cementboard that is coated with baked enamel.

(2)   Pre-cast Concrete Panels

This type of panel is often used as an alternative to metal panels; however, a combination of the two is also possible where the metal panel is applied as a veneer. Generally ceramic tile is cast into the underside of the panel as the final finish.

(3)   Metal Tiles

This tile system is uncommon, but has been used successfully in certain tunnel applications. Metal tiles are coated with porcelain enamel and are set in mortar similarly to ceramic tile.

B. VENTILATION SYSTEMS

1. Types

Tunnel ventilation systems can be categorized into five main types or any combination of these five. The five types are as follows:

- Natural Ventilation
- Longitudinal Ventilation
- Semi-Transverse Ventilation
- Full-Transverse Ventilation
- Single-Point Extraction.

It should be noted that ventilation systems are more applicable to highway tunnels due to high concentration of contaminants. Rail transit tunnels often have ventilation systems in the stations or at intermediate fan shafts, but during normal operations rely mainly on the piston effect of the train pushing air through the tunnel to remove stagnant air. Many rail transit tunnels have emergency mechanical ventilation that only works in the event of a fire. For further information on tunnel ventilation systems refer to NFPA 502 (National Fire Protection Association).

a) Natural Ventilation

A naturally ventilated tunnel is as simple as the name implies. The movement of air is controlled by meteorological conditions and the piston effect created by moving traffic pushing the stale air through the tunnel. This effect is minimized when bi-directional traffic is present. The meteorological conditions include elevation and temperature differences between the two portals, and wind blowing into the tunnel. Figure 2.13 shows a typical profile of a naturally ventilated tunnel. Another configuration would be to add a center shaft that allows for one more portal by which air can enter or exit the tunnel. Many naturally ventilated tunnels over 180 m (600 ft) in length have mechanical fans installed for use during a fire emergency.

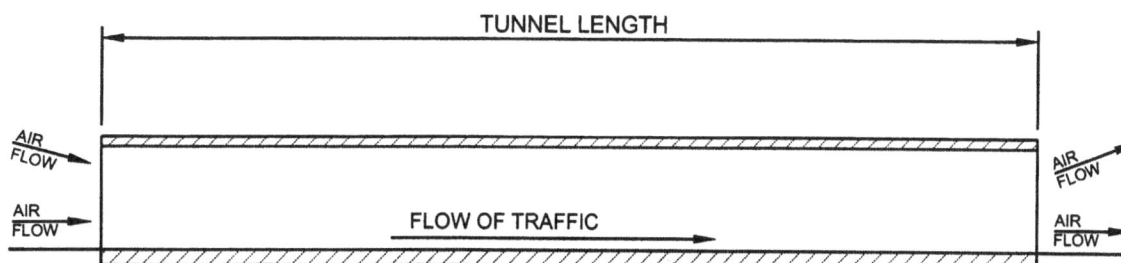

Figure 2.13 – Natural Ventilation

b)   Longitudinal Ventilation

Longitudinal ventilation is similar to natural ventilation with the addition of mechanical fans, either in the portal buildings, the center shaft, or mounted inside the tunnel. Longitudinal ventilation is often used inside rectangular-shaped tunnels that do not have the extra space above the ceiling or below the roadway for ductwork. Also, shorter circular tunnels may use the longitudinal system since there is less air to replace; therefore, the need for even distribution of air through ductwork is not necessary. The fans can be reversible and are used to move air into or out of the tunnel. Figure 2.14 shows two different configurations of longitudinally ventilated tunnels.

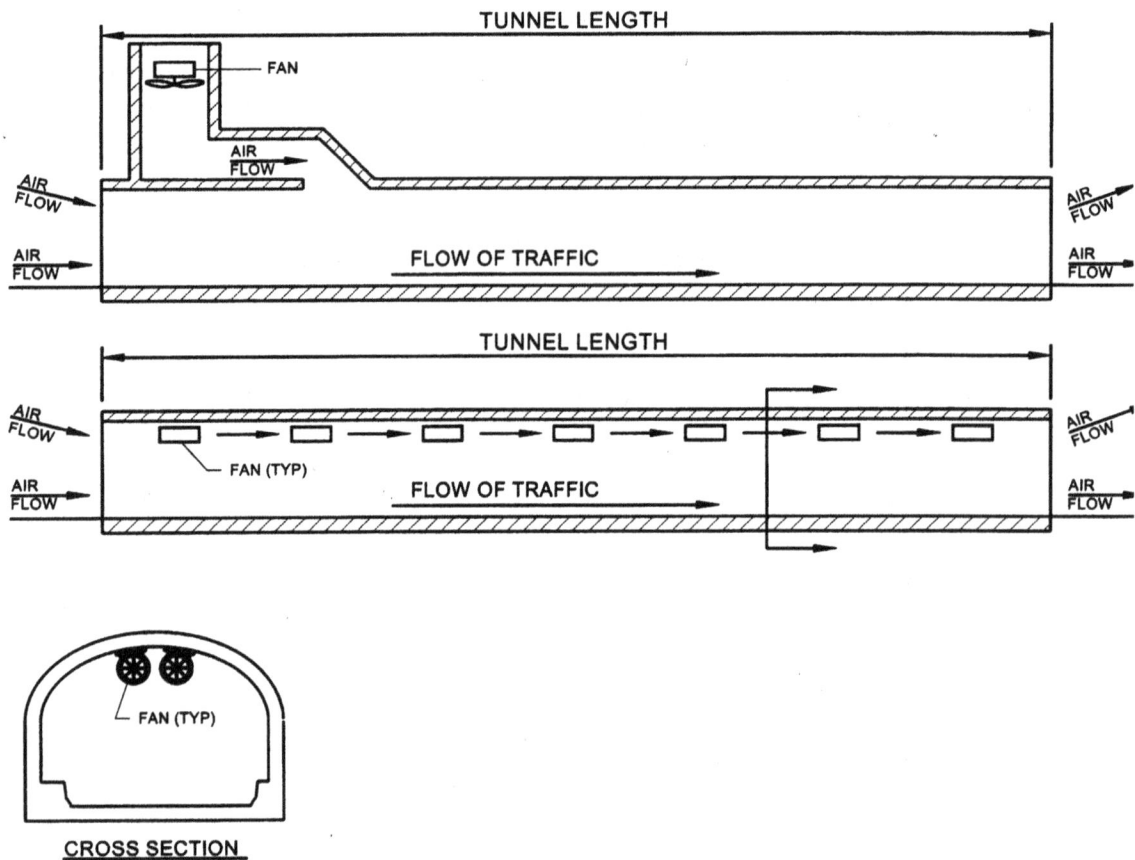

Figure 2.14 – Longitudinal Ventilation

c)   Semi-Transverse Ventilation

Semi-transverse ventilation also makes use of mechanical fans for movement of air, but it does not use the roadway envelope itself as the ductwork. A separate plenum or ductwork is added either above or below the tunnel with flues that allow for uniform distribution of air into or out of the tunnel. This plenum or ductwork is typically located above a suspended ceiling or below a structural slab within a tunnel with a circular cross-section. Figure 2.15 shows one example of a supply-air semi-transverse system and one example of an exhaust-air semi-transverse system. It should be noted that there are many variations of a semi-transverse system. One such variation would be to have half the tunnel be a supply-air system and the other half an exhaust-air system. Another variation is to have supply-air fans housed at both ends of the plenum that push air directly into the plenum, towards the center of the tunnel. One last variation is to have a system that can either be exhaust-air or supply-air by utilizing reversible fans or a louver system in the ductwork that can change the direction of the air. In all cases, air either enters or leaves at both ends of the tunnel (bi-directional traffic flow) or on one end only (uni-directional traffic flow).

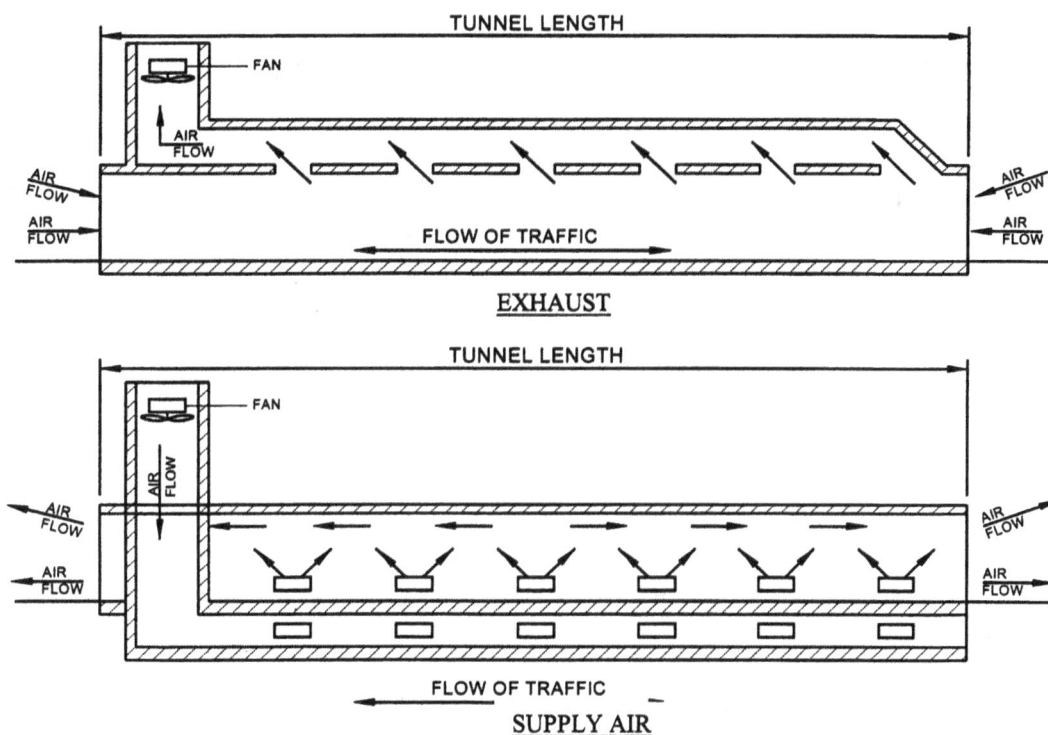

Figure 2.15 – Semi-Transverse Ventilation

d)   Full-Transverse Ventilation

Full-transverse ventilation uses the same components as semi-transverse ventilation, but it incorporates supply air and exhaust air together over the same length of tunnel. This method is used primarily for longer tunnels that have large amounts of air that need to be replaced or for heavily traveled tunnels that produce high levels of contaminants. The presence of supply and exhaust ducts allows for a pressure difference between the roadway and the ceiling; therefore, the air flows transverse to the tunnel length and is circulated more frequently. This system may also incorporate supply or exhaust ductwork along both sides of the tunnel instead of at the top and bottom. Figure 2.16 shows an example of a full-transverse ventilation system.

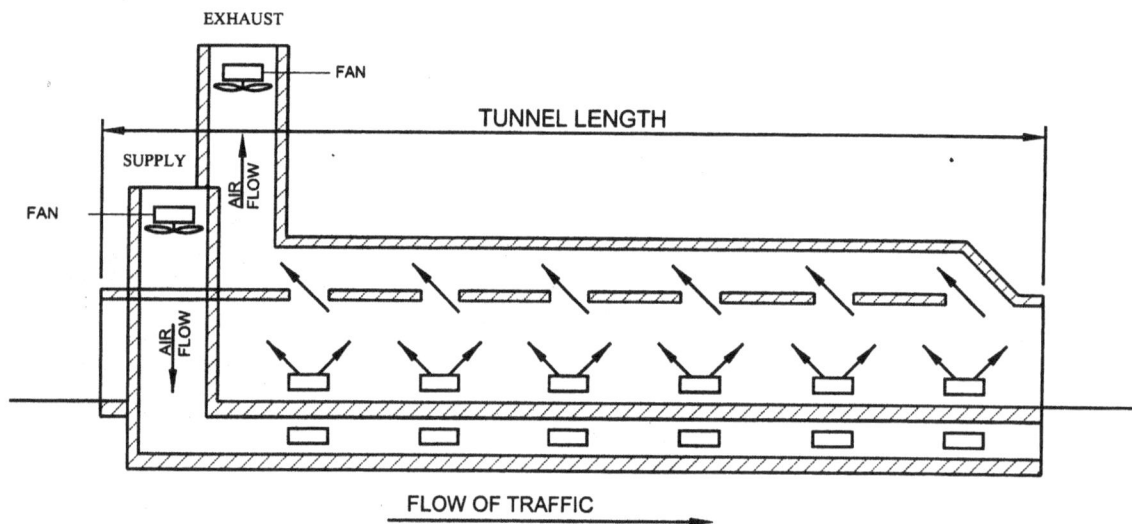

Figure 2.16 – Full-Transverse Ventilation

e)   Single-Point Extraction

In conjunction with semi- and full-transverse ventilation systems, single-point extraction can be used to increase the airflow potential in the event of a fire in the tunnel. The system works by allowing the opening size of select exhaust flues to increase during an emergency. This can be done by mechanically opening louvers or by constructing portions of the ceiling out of material that would go from a solid to a gas during a fire, thus providing a larger opening. Both of these methods are rather costly and thus are seldom used. Newer tunnels achieve equal results simply by providing larger extraction ports at given intervals that are connected to the fans through the ductwork.

2. Equipment

a) Fans

(1) Axial

There are two main types of axial fans—tube axial fans and vane axial fans. Both types move air parallel to the impellor shaft, but the difference between the two is the addition of guide vanes on one or both sides of the impellor for the vane axial fans. These additional vanes allow the fan to deliver pressures that are approximately four times that of a typical tube axial fan. The two most common uses of axial fans are to mount them horizontally on the tunnel ceiling at given intervals along the tunnel or to mount them vertically within a ventilation shaft that exits to the surface.

Tube Axial Fan   Vane Axial Fan

Figure 2.17 – Axial Fans

(2) Centrifugal

This type of fan outlets the air in a direction that is 90 degrees to the direction at which air is obtained. Air enters parallel to the shaft of the blades and exits perpendicular to that. For tunnel applications, centrifugal fans can either be backward-curved or airfoil-bladed. Centrifugal fans are predominantly located within ventilation or portal buildings and are connected to supply or exhaust ductwork. They are commonly selected over axial fans due to their higher efficiency with less horsepower required and are therefore less expensive to operate.

Figure 2.18 – Centrifugal Fan

b)   Supplemental Equipment

(1)   Motors

Electric motors are typically used to drive the fans. They can be operated at either constant or variable speeds depending on the type of motor. According to the National Electric Manufacturers Association (NEMA), motors should be able to withstand a voltage and frequency adjustment of +/- 10 percent.

(2)   Fan Drives

A motor can be connected to the fan either directly or indirectly. Direct drives are where the fan is on the same shaft as the motor. Indirect drives allow for flexibility in motor location and are connected to the impellor shaft by belts, chains, or gears. The type of drive used can also induce speed variability for the ventilation system.

(3)   Sound Attenuators

Some tunnel exhaust systems are located in regions that require the noise generated by the fans to be reduced. This can be achieved by installing cylindrical or rectangular attenuators either mounted directly to the fan or within ductwork along the system.

(4)   Dampers

Objects used to control the flow of air within the ductwork are considered dampers. They are typically used in a full open or full closed position, but can also be operated at some position in between to regulate flow or pressure within the system.

C.   LIGHTING SYSTEMS

1.   Types

a)   Highway Tunnels

There are various light sources that are used in tunnels to make up the tunnel lighting systems. These include fluorescent, high-pressure sodium, low-pressure sodium, metal halide, and pipe lighting, which is a system that may use one of the preceding light source types. Systems are chosen based on their life cycle costs and the amount of light that is required for nighttime and daytime illumination. Shorter tunnels will require less daytime lighting due to the effect of light entering the portals on both ends, whereas longer tunnels will require extensive lighting for both nighttime and daytime conditions. In conjunction with the lighting system, a highly reflective surface on the walls and ceiling, such as tile or metal panels, may be used.

Fluorescent lights typically line the entire roadway tunnel length to provide the appropriate amount of light. At the ends of the roadway tunnel, low-pressure sodium lamps or high-pressure sodium lamps are often combined with the fluorescent lights to provide higher visibility when drivers' eyes are adjusting to the decrease in natural light. The transition length of tunnel required for having a higher lighting capacity varies from tunnel to tunnel and depends on which code the designer uses.

Both high-pressure sodium lamps and metal halide lamps are also typically used to line the entire length of roadway tunnels. In addition, pipe lighting, usually consisting of high-pressure sodium or metal halide lamps and longitudinal acrylic tubes on each side of the lamps, are used to disperse light uniformly along the tunnel length.

b)   Rail Transit Tunnels

Rail transit tunnels are similar to highway tunnels in that they should provide sufficient light for train operators to properly adjust from the bright portal or station conditions to the darker conditions of the tunnel. Therefore, a certain length of brighter lights is necessary at the entrances to the tunnels. The individual tunnel owners usually stipulate the required level of lighting within the tunnel. However, as a minimum, light levels should be of such a magnitude that inspectors or workers at track level could clearly see the track elements without using flashlights.

D. OTHER SYSTEMS/APPURTENANCES

1. Track

The track system contains the following critical components:

a) Rail

The rail is a rolled, steel-shape portion of the track to be laid end-to-end in two parallel lines that the train or vehicle's wheels ride atop.

b) Rail Joints

Rail joints are mechanical fastenings designed to unite the abutting end of contiguous bolted rails.

c) Fasteners/Bolts/Spikes

These fasteners include a spike, bolt, or another mechanical device used to tie the rail to the crossties.

d) Tie Plates

Tie plates are rolled steel plates or a rubberized material designed to protect the timber crosstie from localized damage under the rails by distributing the wheel loads over a larger area. They assist in holding the rails to gage, tilt the rails inward to help counteract the outward thrust of wheel loads, and provide a more desirable positioning of the wheel bearing area on the rail head.

e) Crossties

Crossties are usually sawn solid timber, but may be made of precast reinforced concrete or fiber reinforced plastic. The many functions of a crosstie are to:
- Support vertical rail loads due to train weight.
- Distribute those loads over a wide area of supporting material.
- Hold fasteners that can resist rail rotation due to laterally imposed loads.
- Maintain a fixed distance between the two rails making up a track.
- Help keep the two rails at the correct relative elevation.
- Anchor the rails against both lateral and longitudinal movement by embedment in the ballast.
- Provide a convenient system for adjusting the vertical profile of the track.

f)   Ballast

Ballast is a coarse granular material forming a bed for ties, usually rocks. The ballast is used to transmit and distribute the load of the track and railroad rolling equipment to the subgrade; restrain the track laterally, longitudinally, and vertically under dynamic loads imposed by railroad rolling equipment and thermal stresses exerted by the rails; provide adequate drainage for the track; and maintain proper cross-level surface and alignment.

g)   Plinth Pads

Plinth pads are concrete support pads or pedestals that are fastened directly to the concrete invert. These pads are placed at close intervals and permit the rail to span directly from one pad to another.

2.   Power (Third Rail/Catenary)

a)   Third Rail Power System

A third rail power system will consist of the elements listed below and will typically be arranged as shown in Figures 2.19 and 2.20.

(1)   Steel Contact Rail

Steel contact rail is the rail that carries power for electric rail cars through the tunnel and is placed parallel to the other two standard rails.

(2)   Contact Rail Insulators

Contact rail insulators are made either of porcelain or fiberglass and are to be installed at each supporting bracket location.

(3)   Protection Board

Protection boards are placed above the steel contact rail to "protect" personnel from making direct contact with this rail. These boards are typically made of fiberglass or timber.

(4)   Protection Board Brackets

Protection board brackets are mounted on either timber ties or concrete ties/base and are used to support the protection board at a distance above the steel contact rail.

(5)   Third Rail Insulated Anchor Arms

Third rail insulated anchor arms are located at the midpoint of each long section, with a maximum length for any section limited to 1.6 km (1 mile).

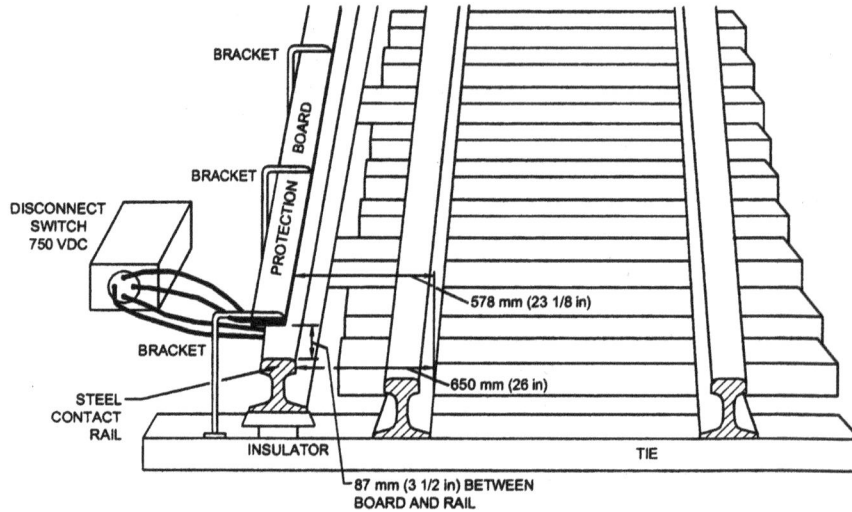

Figure 2.19 – Typical Third Rail Power System
(Note: Dimensions indicate minimum clearance requirements)

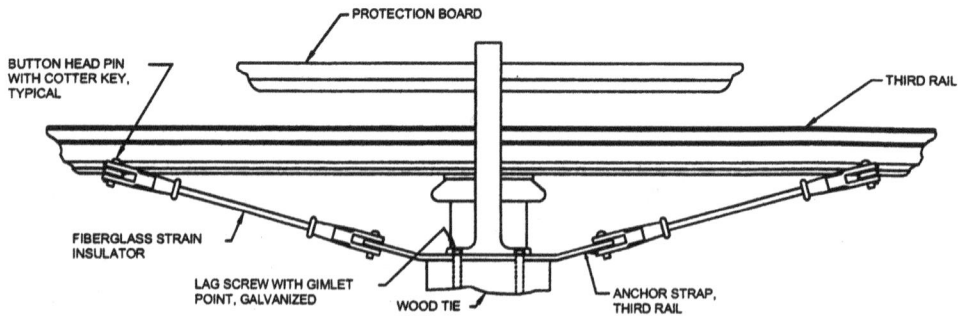

Figure 2.20 – Typical Third Rail Insulated Anchor Arm

b)   Catenary Power System

The catenary system is an overhead power system whereby the rail transit cars are powered by means of contact between the pantographs on top of the rail car and the catenary wire. A typical catenary system may consist of some or all of the following components: balance weights, yoke plates, steady arms, insulators, hangers, jumpers, safety assemblies, pull-off arrangements, back guys and anchors, underbridge assemblies, contact wires, clamped electrical connectors, messenger supports, registration assemblies, overlaps, section insulators, phase breaks, and section disconnects. For tunnel catenary systems, some of the above components are not necessary or are modified in their use. This is particularly true for the methods of support in that the catenary system is supported directly from the tunnel structure instead of from poles with guy wires.

Since the methods used to support a catenary system within a tunnel can vary, a detailed description of the individual components is not given in this section. For inspection purposes, Chapter 4, Section D, Part 2 provides inspection procedures for various components listed above that may exist in a tunnel catenary system.

3.   **Signal/Communication Systems**

a)   Signal System

The signal system is a complex assortment of electrical and mechanical instruments that work together to provide direction for the individual trains within a transit system. A typical signal system may consist of some or all of the following components: signals, signal cases, relay rooms, switch machines, switch circuit controllers, local cables, express cables, signal power cables, duct banks, messenger systems, pull boxes, cable vaults, transformers, disconnects, and local control facilities.

b)   Communication System

The communication system consists of all devices that allow communication from or within a tunnel. Examples of these systems would be emergency phones that are located periodically along a highway tunnel and radios by which train controllers correspond with each other and central operations. The specific components included in a communication system include the phones and radios, as well as any cables, wires, or other equipment that is needed to transport the messages.

CHAPTER 3:
PREVENTIVE MAINTENANCE

A.   PREVENTIVE MAINTENANCE OF THE TUNNEL STRUCTURE

The primary objectives of incorporating regular preventive maintenance procedures into the tunnel structure and its systems are to provide a safe and functional environment for those who work in or travel through the tunnel and to extend its useful life. Since it is usually not possible to have advance knowledge of where structural defects will occur, it is important that regular in-depth inspections be performed in which structural defects are identified and subsequently scheduled for repair based upon their severity. Chapter 4 deals with methods for repairing such structural defects. Aside from predicting structural defects, there are other preventive maintenance tasks that can be performed regularly to ensure safe operation of the tunnel. These tasks include:

- Tunnel Washing
- Drain Flushing
- Ice/Snow Removal
- Tile Removal.

A description of each maintenance task is provided below.

1.   Tunnel Washing

It is recommended that highway tunnels that utilize an interior finish, such as ceramic tile, porcelain enameled panels, etc., be washed according to the following procedure: first, spray tunnel with water or a water/detergent mixture if permitted and scrub with mechanically rotating brushes; second, rinse tunnel with water using high-pressure jets. The primary reason for performing tunnel washing is to maintain proper tunnel luminance, which is dependent on the reflectivity of the tunnel finish. Highway tunnels with unfinished surfaces (bare concrete or exposed rock) and rail transit tunnels do not typically require washings, because reflectivity of the surface is not critical.

The frequency of this procedure may vary for each tunnel owner and depends on environmental conditions. It is recommended that washings be suspended during winter months for tunnels that are located in a region where wintertime temperatures are below freezing. Another factor in determining frequency would be the average daily traffic (ADT) that uses the tunnel. Since most of the dirt is from vehicle exhaust and tire spray, tunnels with a lower ADT would not accumulate dirt as quickly and can be washed less frequently.

2.   Drain Flushing

Roadway drain inlets or drainage troughs in the case of direct fixation track should be kept free of debris and should be flushed with water to verify that drains are operating correctly. This should be done on a semi-annual basis. For highway tunnels, it can be performed concurrently with tunnel washing since the flushing equipment will be available.

3. Ice/Snow Removal

In regions where the temperature within the tunnel drops below freezing, ice forms at locations of active leakage. When such ice could build up on the roadway or safety walk, it is critical that deicing agents be used to prevent accumulation of ice that could present a danger to automobile traffic or tunnel personnel using the safety walk. During these potential icing conditions, it is suggested that the tunnel be inspected daily to observe and to take action to mitigate such leakage.

Also, in similar regions where snow and ice may accumulate for a certain distance within the tunnel from the portals, it is essential that proper plowing be performed and deicing agents be applied to maintain safe traveling conditions. As can be expected, the frequency of such a task is dependent on the natural conditions that produce the snow and ice.

4. Tile Removal

During an in-depth inspection, areas of loose tiles should be identified and those that are in danger of falling should be removed. It is recommended that those loose tiles, which remain, be inspected on a quarterly basis to determine if more tiles need to be removed to ensure safety to the motorist. Another time of identifying and removing possible loose tiles is during the monthly tunnel washing procedure. Often, tiles will become dislodged during the scrubbing or pressure washing of the tunnel. Any new areas should be noted and added to the list of areas to be inspected on a quarterly basis. Any tiles that are removed should be scheduled for replacement.

B. PREVENTIVE MAINTENANCE OF MECHANICAL SYSTEMS

The tunnel mechanical systems are comprised of multiple individual components, many of which must work together for the overall systems to function properly. Since these overall systems are critical for providing a safe environment for the tunnel users and staff, it is paramount that they be well maintained to prevent unforeseen breakdowns. To achieve this goal, it is recommended that a routine preventive maintenance program be developed that includes every major piece of equipment and that work orders be generated on a set schedule for the tasks that are to be performed. To assist in this process, multiple computerized database systems have been developed that can be adapted to a particular tunnel owner's needs. If a computerized database system is used, it would have the capability of storing historical repair, replacement, and cost data for use in properly predicting the life-cycle costs for a particular piece of equipment.

It is impossible for the scope of this manual to incorporate preventive maintenance procedures for every conceivable piece of equipment; however, the major components of the mechanical systems are included. Many tunnels may not utilize all of the components listed due to their size, location, or age; whereas, newer tunnels and tunnels yet to be built may incorporate new technologies that to date have not been addressed. For this reason, it is always necessary to follow the manufacturers' suggested preventive maintenance procedures for a given piece of equipment, particularly if it differs from that given below.

Also, it should be noted that the preventive maintenance functions given are sometimes general and therefore should be made specific to the actual equipment that exists in a particular tunnel. Table 3.1 lists the preventive maintenance functions for each of the major pieces of equipment or mechanical systems along with the suggested frequency for performing the preventive maintenance.

Table 3.1 – Preventive Maintenance of Mechanical Systems

Procedure Description	Weekly	Monthly	Bi-Monthly	Quarterly	Semi-Annually	Annually	Bi-Annually	Tri-Annually
Air Compressor								
Clean or replace air filters if necessary				X				
Clean external cooling fans				X				
Manually operate safety valves and drain tank				X				
Inspect oil for contamination and change if necessary						X		
Check belt tension, clean motor, and operate safety valves on receiver						X		
Inspect for air leaks						X		
Tighten or check all bolts and lubricate motor bearings						X		
Inspect and clean compressor valves						X		
Verify operation of low-level oil switch						X		
Check all pressure and safety controls						X		
Air Conditioning Unit								
Clean or replace air filters		X						
Check coils and clean if necessary						X		
Inspect controls and verify proper operation of unit						X		
Boilers (Furnaces)								
Check chimney and flue for obstructions and make sure all joints are well supported and properly sealed						X		
Lubricate pumps and motors as required						X		
Clean entire boiler, inside and out						X		
Replace fuel filter and oil atomizing nozzle						X		
Check hot water levels and fill as necessary						X		
Restart boiler and test burner performance, flue gas CO_2, smoke, and temperature						X		
Verify operation of all limit switches and primary controls						X		
Test relief valve or safety valve (use extreme caution)						X		
Chiller								
Check and lubricate compressors					X			
Check safety controls						X		
Clean and inspect barrel						X		
Check and add chemicals (as indicated or as required)						X		
CO Monitoring Equipment								
Local Sensors (Calibration and/or sensor replacement)					X			
Vacuum Tubing (Leak Test)						X		
Vacuum Pump (lubrication)				X				

Federal Transit Administration

Procedure Description	Weekly	Monthly	Bi-Monthly	Quarterly	Semi-Annually	Annually	Bi-Annually	Tri-Annually
Frequency (header spanning)								
Central Sensor Calibration (as required by individual system)					X			
Comparison Gas Refill (as required)		X						
Cooling Towers								
Check and lubricate pumps and fans					X			
Check safety controls						X		
Clean sump						X		
Check and add chemicals (as indicated or as required)						X		
Domestic Water Pump and Tank								
Visually inspect pump (when accessible)		X						
Lubricate pump and motor						X		
Check pump operation in conjunction with well tanks						X		
Lubricate ejector pumps						X		
Measure water drawdown to verify proper operation						X		
Check air pressure in tank and correct as necessary						X		
Verify start and stop settings of pressure switch (differential should not exceed 172 kPa (25 psi))						X		
Drainage System								
Grate inspection	X							
Flushing of inlet and piping system					X			
Dewatering Pumps (Fixed and Portable)								
Clean and visually inspect					X			
Lubricate pumps (prior to use for portable)						X		
Emergency Eyewash								
If bacteria control solution is not used, flush and clean unit with pure water				X				
Drain unit and flush and clean the storage tank and refill with water and water treatment				X				
Exhaust Fans and Dampers (Not Tunnel Fans)								
Operate fans and motor operated dampers and listen for unusual noises and vibrations		X						
Check bearings and inspect V-belts for tightness						X		
Clean centrifugal wheel, inlet, and other moving parts						X		
Lubricate shaft bearing pillow blocks and motor bearings						X		
Fire Extinguishers								
Inspect each fire extinguisher in the tunnel				X				
Fire Hydrants								
Grease top nut						X		

Procedure Description	Frequency							
	Weekly	Monthly	Bi-Monthly	Quarterly	Semi-Annually	Annually	Bi-Annually	Tri-Annually
Fire Lines								
Freeze Protection Pumps								
Clean and visually inspect				X				
Lubricate and grease pumps						X		
Heat Tracing Equipment								
Verify system operation (prior to system operation)						X		
Fire Pumps								
Visually inspect pump		X						
Operate pump				X				
Lubricate pump, motor, and coupling					X			
Operate pump and measure current					X			
Check shaft endplay					X			
Check and correct pressure gauges as required					X			
Fire Pump Controller								
Exercise isolating switch and circuit breaker		X						
Operate pumps from both alternate and primary power supplies		X						
Conduct annual test of system including flow and no flow conditions in accordance with NFPA 72						X		
Fire Tank Fill Pump								
Visually inspect pump		X						
Lubricate pump and motor						X		
Fuel Oil Day Tank								
Inspect tank for damage, corrosion, or leakage on both inside and outside of tank. Perform during same week as boiler inspection						X		
Hot Water Pump								
Visually inspect plumbing connections for signs of corrosion					X			
Visually inspect exterior of water heater for signs of leakage					X			
Lubricate pump and motor as required						X		
Septic System								
Pump out tank (as indicated or as required)							X	
Ejector Pumps								
Check local indications (verification of proper functioning from control panel)	X							
Visually inspect pumps						X		
Tunnel Fans								
Check motor bearings	X							
Listen for any unusual noise or vibration	X							
General cleaning of motor, interior and exterior			X					

Procedure Description	Frequency							
	Weekly	Monthly	Bi-Monthly	Quarterly	Semi-Annually	Annually	Bi-Annually	Tri-Annually
Disconnect motor from power supply and regrease, making sure chamber is 75 percent full of grease				X				
Operate fan through entire range of speeds and note any noises or vibrations (Balance fan if required)				X				
Inspect inside and outside of housing and impellor for wear, deterioration, or build-up of material				X				
Inspect mounting bolts, anchors, and connections for failures or damage				X				
Change oil in pillow blocks and drive guards (mineral oil is recommended)				X				
Remove inspection cover from drive guard and inspect chain to verify proper lubrication and wear and adjust if necessary				X				
Check all oils and greases for contaminants				X				
Verify that any dampers operate properly through all positions, and lubricate if necessary				X				
Unit Heaters								
Clean unit casing, fan, diffuser, coil, and/or motor thoroughly, and clean and repaint any corrosion spots on casing						X		
Tighten the fan guard, motor frame, and fan bolts, and check fan clearances						X		
Inspect any control panel wiring to ensure that the insulation is intact and that all connections are tight						X		
Examine all heater and relay contacts for pitting or burning and replace if necessary						X		
Lubricate motor if necessary						X		
Check operation of hydronic controls								
Underground Fuel Oil Tank								
Remove liquid level sensor from reservoir to check low-level alarm				X				
Immerse sensor into bucket of water to activate high-level alarm				X				
Water Storage Tank								
Visually inspect tank exterior				X				
Drain sediment					X			
Observe water system operation and note any abnormal happenings						X		
Measure water draw down to verify proper operation						X		
Check air pressure in tank and correct as necessary						X		
Verify start and stop settings of pressure switch (differential should not exceed 172 kPa (25 psi))						X		
Visually inspect tank interior								X

C. PREVENTIVE MAINTENANCE OF ELECTRICAL ELEMENTS

Similar to tunnel mechanical systems, many individual components make up the tunnel electrical systems. However, one difference is that many of the electrical components are interconnected and rely on each other for proper functioning of the entire system. Also, the electrical systems could be viewed with higher importance because the mechanical systems and other tunnel systems need electricity for them to function properly. Given the importance of an electrical system that is constantly being used and is vital for the overall safety of the tunnel, it is suggested that the preventive maintenance system that was recommended for the mechanical systems be adapted to include preventive maintenance functions for the electrical systems.

As with the mechanical systems, only the major components of the electrical systems are included herein. Many tunnels may not utilize all of the components listed due to their size, location, or age; whereas, newer tunnels and tunnels yet to be built may incorporate new technologies that to date have not been addressed. For this reason, it is always necessary to refer to the manufacturers' suggested preventive maintenance procedures for a given piece of equipment. Additionally, the InterNational Electrical Testing Association (NETA), in their Maintenance Testing Specifications (MTS-2001), provides detailed information and guidelines regarding maintenance of electrical equipment. More specifically, Appendix B of MTS-2001 provides recommended frequencies for maintenance procedures that are comparable to what is given in this section. Another reference is the National Fire Protection Association's *NFPA 70B: Recommended Practice for Electrical Equipment Maintenance*.

For the procedures given below to be performed efficiently and safely, it is recommended that in-house maintenance staff be trained in the current Occupational Safety and Health Administration (OSHA) and NFPA standards, including but not limited to *NFPA 70E: Standard for Electrical Safety Requirements for Employee Workplaces*. If the tunnel owner does not have qualified in-house personnel, it is recommended that an outside electrical testing agency be contracted that meets the requirements of NETA full membership. Also, a switching procedure and one-line safety diagrams of the electrical system should be prepared and posted in all electrical rooms.

As with the mechanical preventive maintenance functions, the electrical preventive maintenance functions given are sometimes general and should be made specific to the actual equipment that exists in a particular tunnel. Table 3.2 lists the preventive maintenance functions for each of the major pieces of equipment or electrical systems along with the suggested frequency for performing the preventive maintenance.

Table 3.2 – Preventive Maintenance of Electrical Systems

Procedure Description	Frequency							
	Weekly	Monthly	Bi-Monthly	Quarterly	Semi-Annually	Annually	Bi-Annually	Tri-Annually
Closed Circuit TV								
Clean, align, and focus all cameras after tunnel washing					X			
Emergency Lighting								
Operate test buttons on emergency light fixtures		X						
Operate battery pack for emergency lighting for 90 minutes						X		
Electrical Switchboard and Switchgear								
Inspect switchgear bus and connections by infrared scanning						X		
Perform ultrasonic inspection of medium voltage switchgear bus supports, insulators, and barriers						X		
Visually inspect all equipment for unusual conditions						X		
Check tightness of all connections						X		
Remove and replace defective lighting contacts						X		
Review results of last visual, infrared, and ultrasonic inspection								X
After power shutdown, clean entire switchgear interior								X
Clean all bus insulators and check for cracks and chips								X
Clean, lubricate (if applicable), and verify operation of all control switches, auxiliary relays, and devices								X
Clean, lubricate, adjust, and add anti-oxidant grease to contacts of all disconnect switches								X
Clean and perform insulation resistance testing on all lightning arrestors								X
Perform insulation resistance testing on any bus bars								X
Perform calibration test and verify proper operation of all meters								X
Low Voltage Air Circuit Breakers								
Remove covers and thoroughly clean each breaker and contact surfaces								X
Apply anti-oxidant grease to breaker's main contacts								X
Lubricate and verify operation of all mechanisms								X
Apply current equal to 90 to 110 percent of the breaker trip coil setting to verify proper pick-up of tripping mechanism								X
Record trip times for long-time, short-time instantaneous, and ground fault breakers when passing loads equal to multiples of their listed ratings through each phase of the breaker								X
Measure contact resistance and adjust where possible								X
Perform and record results of insulation resistance test from each pole to other two poles and to ground								X

Procedure Description	Frequency							
	Weekly	Monthly	Bi-Monthly	Quarterly	Semi-Annually	Annually	Bi-Annually	Tri-Annually
Clean and lubricate breaker carriage and racking mechanism on any draw out breakers								X
Molded Case Circuit Breakers								
Inspect breaker for proper installation								X
Remove cover (if possible) and fully clean interior and exterior								X
Inspect for burning, overheating, wear, and proper alignment								X
Perform contact resistance and insulation resistance measurements and test								X
Apply current equal to 300 percent of breaker rating to test the long-time element								X
Test and compare any breakers with instantaneous trip units to manufacturer's characteristic curve								X
Automatic Transfer Switch (600 Volt Class)								
After total outage is obtained, clean all contact surfaces, apply anti-oxidant contact grease, measure and record contact resistance, and make any adjustments if necessary								X
Lubricate bearings, links, pins, and cams								X
Perform insulation resistance test								X
Test all settings of voltage, frequency sensing, and timing relays								X
Low Voltage Insulated Cable (Less Than 600 Volts)								
Check all cable terminations for tightness								X
Perform and record results of insulation resistance test from each phase to the other two and to ground for one minute using a test voltage of 1,000 volts Direct Current (DC). Compare results with previous tests.								X
Electrical Transformer								
Inspect transformer connections by infrared scanning						X		
Perform ultrasonic inspection of medium voltage bus supports, insulators, and barriers						X		
Visually inspect all equipment for unusual conditions						X		
Test transformer and circuit breaker insulating oil						X		
Dry-Type								
Remove cover and visually inspect all cable/bus connections for evidence of overheating or burning, check for tightness and clean windings								X
Liquid-Filled								
Inspect transformers for leaks, deteriorated seals/gaskets, proper oil level, and test oil sample								X

Procedure Description	Weekly	Monthly	Bi-Monthly	Quarterly	Semi-Annually	Annually	Bi-Annually	Tri-Annually
Inspect transformer tank and cooling fins for corrosion, chipped paint, dents, and proper connection to ground								X
Inspect all bushings for cracks/chips, proper tightness, and evidence of overheating								X
Inspect all gauges and alarm devices								X
Clean core, coils, and enclosures and inspect any filters								X
Perform primary and secondary insulation resistance test where possible.								X
Perform polarization index test on transformers 500 KVA and larger								X
Perform turns ratio tests								X
Perform calibration test and verify proper operation of all meters								X
Fire Alarm System								
Perform all tests and inspections in accordance with NFPA 72								
Make and file a permanent record of all inspections and tests conducted								
Open primary power supply to fire alarm panel and note sounding of trouble alarm and light		X						
Perform fire drill by use of drill switch on fire alarm panels, and check that all visual and audible signals emit a sound and tunnel SCADA system (if any) receives alarm		X						
Visually inspect all supervisory and water flow alarms on any standpipe systems		X						
Test all heat detectors with a calibrated heat source and replace all failed units					X			
Test all smoke detectors by measuring and recording sensitivity; replace all failed units					X			
Clean all smoke and heat detector housings and check battery voltage under load					X			
Verify that proper alarm devices operate for the appropriate initiating device circuit					X			
Verify that all remote annunciators operate				X				
Check all lamps, alarm devices, and printers for proper operation				X				
Make a discharge test of batteries to determine capacity for operating system for 24 hours					X			
Generator								
Operate unit under load for 4 hours and check lubrication levels		X						
Change oil, coolant, and filter				X				

Procedure Description	Weekly	Monthly	Bi-Monthly	Quarterly	Semi-Annually	Annually	Bi-Annually	Tri-Annually
					Frequency			
Compare nameplate information and connection with drawings and specifications				X				
Inspect for proper anchorage and grounding				X				
Perform insulation resistance test on generator winding with respect to ground and determine polarization index				X				
Perform phase rotation test to determine compatibility with load requirements				X				
Functionally test engine shutdown and alarm controls for low oil pressure, overtemperature, overspeed, and other features				X				
Perform vibration base-line test and plot amplitude versus frequency for each main bearing cap				X				
Perform load bank test and record voltage, frequency, load current, oil pressure, and coolant temperature at periodic intervals during test				X				
Monitor and verify correct operation and timing of normal voltage-sensing relays, engine start sequence, time delay upon transfer, alternate voltage-sensing relays, automatic transfer operation, interlocks, limit switch functions, time delay and retransfer upon normal power restoration, and engine cool down and shutdown feature				X				
High Voltage Disconnect								
Inspect disconnect switch bus and connections by infrared scanning						X		
Perform ultrasonic inspection of medium voltage bus supports, insulators, and barriers						X		
Visually inspect all equipment for unusual conditions						X		
Busing Inspection								
Review results of last visual, infrared, and ultrasonic inspection								X
Check for proper tightness of all exposed bus connections								X
Thoroughly clean and check for cracks/chips of all bus insulators								X
Clean, lubricate (if applicable), and verify operation of all control switches, auxiliary relays, and devices								X
Clean, lubricate, adjust, and add anti-oxidant grease to contacts of all disconnect switches								X
Clean and perform insulation resistance test on all lightning arrestors								X
Perform insulation resistance test on any bus bars								X
Service Enclosed Air Break Switches								
After shutdown, clean and inspect entire switch mechanism								X

Procedure Description	Frequency							
	Weekly	Monthly	Bi-Monthly	Quarterly	Semi-Annually	Annually	Bi-Annually	Tri-Annually
Check switch contacts for proper alignment and apply anti-oxidant grease to main contacts								X
Check switch's arcing contacts for proper opening sequence relative to main contacts								X
Inspect fuses and record size and type used								X
Clean all phase isolation barriers and check for contamination and corona damage								X
Thoroughly clean and check for cracks/chips of all insulators								X
Clean and perform insulation resistance test on all lightning arrestors								X
Inspect all ground connections								X
Perform contact resistance and insulation resistance tests and record results								X
Motor Control Center								
Inspect controller bus and connections by infrared scanning						X		
Perform ultrasonic inspection of medium voltage bus supports, insulators, and barriers						X		
Visually inspect all equipment for unusual conditions						X		
Review results of last visual, infrared, and ultrasonic inspections								X
After power shutdown, clean entire controller interior								X
Check for proper tightness of all exposed bus connections								X
Clean all bus insulators and check for cracks and chips								X
Clean, lubricate (if applicable), and verify operation of all control switches, auxiliary relays, and devices								X
Clean, lubricate, adjust, and add anti-oxidant grease to contacts of all disconnect switches								X
Perform an insulating resistance and polarization test of the bus and the motor feeder with the motor connected								X
Test overloads at 125 percent and 600 percent of rating against the tripping curve								X
Perform calibration test and verify proper operation of all meters								X
Lighting Relays and Contactors								
Clean all contacts and replace all worn and pitted contacts								X
Check tightness of contactors								X
Measure load current and verify proper operation								X
Traffic Signals								
Inspect and verify operation of Lane Control Devices		X						
Inspect and verify operation of Variable Message Signs		X						

Procedure Description	Frequency							
	Weekly	Monthly	Bi-Monthly	Quarterly	Semi-Annually	Annually	Bi-Annually	Tri-Annually
Tunnel Control System								
Check all controls on consoles for proper operation of tunnel lighting and fans						X		
Test all alarm and lights for proper feedback from devices						X		
Check all connections for tightness						X		
Clean cabinets						X		
Tunnel Lights								
Verify proper operation of the lighting fixtures in the tunnel areas	X							
Count and record number of lights out on night lighting and day lighting	X							
Replace any inoperable bulbs or ballasts with similar or increased efficiency	X							
Clean exterior of lenses on all lighting fixtures in the tunnel				X				
If required clean interior of lenses				X				
Perform group relamping for specific lamp types						X		
Underground Tank and Piping Monitor								
Perform built-in test (if any) and verify that each circuit is operational. If not, identify circuit using troubleshooting guide and replace parts as necessary						X		

D. PREVENTIVE MAINTENANCE OF TRACK SYSTEMS

1. Track and Supporting Structure

The track and its supporting structure should be inspected more frequently than other systems within a tunnel. In fact, the tasks of inspection and preventive maintenance may often overlap in order to make efficient use of the inspection and maintenance staff and equipment. This does not detract from the importance of proper documentation of the inspection process; it just allows for certain simple procedures to be performed immediately after the condition is discovered. This serves as a means of preventing any further degradation that could occur before a scheduled maintenance is performed. If items are going to be repaired or replaced, it is important that some internal guidelines be followed to ensure accuracy and consistency of repairs or new installations. In lieu of such guidelines it is recommended that the USDOT's Federal Railroad Administration – Office of Safety's Code of Federal Regulations for Title 49, *Track Safety Standards Part 213 Subpart A to F, Class of Track 1-5 (TSS Part 213)* be used.

Listed below are several preventive maintenance procedures that are recommended to prolong the working life of the track and the supporting structure.

a)   Rail Lubrication

It is commonly known that periodic lubrication of curves can extend rail life. The lubricant should be placed on the gage face of the rail, with care taken to minimize the amount of lubricant to prevent migration to the top of the rail head. The application can be performed by hand, by the use of wayside lubricators (a train actuated device that first applies the lubricant to the wheel flanges and then subsequently to the rail), or by railcar-mounted lubricant sticks that apply a thin coat of grease to the gage face during train operation. The frequency of this procedure is based on durability or life expectancy of the lubricant used and the amount of train traffic to which the rail is subjected. The procedure can also be performed if excessive wear is identified during a routine inspection or if noise abatement is desired. Additionally, asphalt based dipping oil should be applied to tie plates and spikes when they are subjected to corrosive conditions. This oil can be applied using a spray machine.

b)   Rail Grinding

In addition to removing defects that are identified using specialized rail defect detection equipment, performing scheduled "out-of-face" grinding and profile grinding of the rail head can help prevent the development of surface defects by optimizing the rail-wheel interaction. The frequency of this procedure is dependent upon the amount of gross tonnage traveling over the track and can range from one year for track with very high tonnages to five years for track with low tonnages.

c) Ballast Cleaning/Replacement

Within a tunnel, the ballast is not subjected to the sedimentation of fine particles within the voids due to excessive vegetation growth; however, if severe water infiltration exists, the ballast can be negatively impacted in a localized area. If known areas exist where the ballast is either being eroded or undermined by water flow, or being fouled with silt carried by the water infiltration, certain tasks can be periodically performed in lieu of addressing the water infiltration problem using methods given in Chapter 4. The ballast can be removed and cleaned using a ballast cleaner and subsequently replaced, or a new layer of ballast (track surfacing) can be applied to the affected area and tamped to match the specified cross section. The entire ballast section along the tunnel should be maintained at all times.

d) Tie Renewals

A routine program of replacing crossties that do not meet inspection standards should be implemented to ensure that the proper number of quality crossties are located within each length of rail.

e) Joint Maintenance

All joints should be fully bolted and the bolts should be retightened as required within a range of 9,070 to 13,610 kg (20,000 to 30,000 lb) per bolt for the initial tightening of a new bolt and between 6,800 to 11,400 kg (15,000 and 25,000 lb) per bolt for all subsequent retightening. It is recommended that the initial retightening be performed one to three months after installation and all subsequent retightening be done on an annual basis. Also, if initial petrolatum or petrolatum-based compound for preserving the joint is deficient, then a new coating should be applied. The spray method can be used so that the integrity of the joint is not disturbed.

f) Regaging

As gage deficiencies are identified during inspection, regaging should be performed if changes in gage are severe or abrupt.

g) General Aligning

Independent of whether the track structure is direct fixation or ballasted construction, the general vertical and horizontal alignment should be periodically adjusted to conform to specified standards. This can be accomplished by using automatic track aligning equipment for ballasted track or by manually raising and lining direct fixation track and placing shims under the rail plates as necessary.

h)   Spike Replacement

If it is suspected that stray current corrosion is occurring in a tunnel, the spike will most likely need to be replaced on a routine basis. The normal life expectancy of 25 years for spikes can be as little as 6 months if stray current corrosion is occurring in combination with the presence of moisture.

2.   **Power (Third Rail/Catenary)**

a)   Third Rail System

The proper operation and efficiency of third rail power systems is crucial to those tunnel track segments that contain them. Therefore, it is beneficial to perform routine inspections of the systems as outlined in the complementary Tunnel Inspection Manual and also to conduct routine preventive maintenance of certain elements that make up the third rail system. Possible procedures to accomplish this are listed below.

- Perform general aligning on third rail to ensure consistency with the running rails alignment.
- Periodically clean rail insulators to prevent stray current from entering the ground or supporting structures and increasing the amount of corrosion. This is especially true in wet environments near portals or areas of water infiltration within the tunnel since moisture also advances the onset of corrosion.
- Repair/replace deficient protection boards and brackets to ensure that they do not interfere with connections to the train or fail to provide safety to tunnel personnel.
- Repair/replace splices and joints that could be impeding the current flow for the contact rail or redirecting flow causing stray current corrosion.

b)   Catenary System

Similar to the third rail power systems, catenary systems are crucial for the proper operation of the transit systems that utilize them. For that reason, it is necessary to perform regular preventive maintenance in addition to the periodic visual and in-depth inspections that are presented in the complementary Tunnel Inspection Manual. Apart from major repairs or complex preventive maintenance tasks, many of the suggested procedures below can be performed at the same time as the in-depth inspections in order to minimize disruption to the system's schedule.

- Replace broken, chipped, or otherwise deficient sheds on all insulators.

- Align hangers to vertical position and rectify condition that may have caused hangers to be out of alignment.
- Replace segments of contact wire with vertical thickness less than 10.7 mm (0.42 in).
- Remove, clean, and tighten "C" jumpers, feeder points, and full section overlap jumpers that have signs of corrosion or burning. Apply high melting point grease to all stranded conductors.
- Adjust contact wires at overlaps for proper matching alignment.
- Adjust turnbuckles on hangers of section insulators to keep units level.

3. Signal/Communication Systems

Signal/communication systems relate directly to the overall safety of the rail transit system. As ridership and train frequencies increase, so does the dependence on a reliable efficient method for maneuvering the cars through the system and for communicating with them during that process or during emergencies. Since a majority of the components that make up these systems are electric or electronic, their proper operation is or can be tested continuously. However, there are mechanical devices that are operated manually or by electric power that should be consistently maintained. Most problems with this equipment are identified during a routine inspection and thus can be fixed immediately or scheduled for immediate action. On the other hand, it is recommended that a routine program be implemented to lubricate moving components, clear debris from path of moving components, and replace light bulbs in crucial equipment.

E. PREVENTIVE MAINTENANCE OF MISCELLANEOUS APPURTENANCES

1. Corrosion Protection Systems

Corrosion protection systems may be used in either highway or rail transit tunnels. Two types of corrosion protection exist: cathodic protection and stray current protection. A description of each of these systems is provided below.

a) Cathodic Protection Systems

Cathodic protection systems are designed to protect any metal components of the tunnel structure or other systems (such as buried pipelines, surrounding buildings, or the rail system itself) from deteriorating prematurely due to corrosion resulting from the presence of any aqueous electrolyte. Corrosion from electrolysis is specifically prevalent in areas where moisture is present and where there are dissimilar metals attached together.

These systems may be as simple as providing connections between metal components and the ground so that electrolysis does not take place in the criticalmetal components. Also, passive cathodic protection systems with anodes buried in the ground, can be used to sacrificially attract the stray current away from the critical metal components. Another method is to counterbalance the effects of stray current by inducing an impressed current using rectifiers.

This section recommends maintenance procedures for effective and efficient operation of cathodic protection systems. As with other systems, specific cathodic protection components will vary from one tunnel to another depending on the how the original design provided protection. Therefore, for complex electrical components such as rectifiers, the manufacturer's recommended maintenance procedures will always take precedence over any recommendations given in this section. Any testing of the cathodic protection effectiveness shall be performed in accordance with the National Association of Corrosion Engineers (NACE) Internationals' recommended practices and procedures. Also, ensure that any test equipment is in good operating condition and that the calibration effective period has not expired.

Perform electrical measurements (voltage and current) and inspection of a cathodic protection system annually. The electrical measurements and inspection will be performed to:

- Make certain that protection is being provided in accordance with established criteria in the design documents.
- Make required adjustment to accommodate changes.
- Locate areas of inadequate protection levels.
- Identify areas that may be affected by future or ongoing construction.
- Adjust frequency of test and inspections to reflect changes to field conditions, safety, and economic considerations.
- Select areas to be monitored more frequently.
- Provide additional equipment as needed to maintain an effective cathodic protection system.
- Assess the effectiveness of isolated joints and continuity jumpers to achieve proper isolation.

Perform semi-annual inspection and testing of the following cathodic protection equipment:

- All impressed current sources (rectifiers and power supplies).
- All impressed current protective devices (protective relays, circuit breakers, fuses, wiring, and lightning protection).
- Reverse current switches, diodes, fuses, and wiring.

Other remedial tasks that may be performed on the cathodic protection system could include the following:

- Replace anodes per manufacturers' criteria.
- Repair or replace any defective components of the cathodic protection system.
- Clean and coat as required to provide isolation.
- Repair or replace jumpers.
- Replace defective wiring.
- Remove any accidental metallic contact.
- Repair or replace any defective isolating device.

Sufficient testing should be performed following any adjustment to the cathodic protection system to assure proper protection to the tunnel structure and that no adverse effects will occur to other nearby structures.

b.   Stray Current Protection Systems

Stray current occurs within DC electrified rail transit systems and in conjunction with moisture from water infiltration, can cause significant amounts of corrosion of tunnel and rail components. The risk of stray current corrosion can be reduced by increasing the resistance of the leakage path to the earth through increasing rail insulation, and reducing the amount of moisture present in the tunnel by addressing the water infiltration problem (as described in Chapter 4, Section A). Aside from the above measures, most new DC traction power systems in tunnels are designed to minimize leakage of DC stray currents. However, older tunnels may require that the DC systems be modified or retrofitted to eliminate corrosion resulting from DC stray currents.

2.   Safety Walks, Railings, and Exit Stairs/Ladders

It is important to ensure that tunnel elements such as safety walks, railings and exit stairs/ladders that are accessed either by tunnel personnel or tunnel occupants in the event of an emergency, be properly maintained such that they will be able to support the pedestrian loadings during their use. This can be accomplished by providing the following preventive maintenance functions:

- Keep areas clean and free of debris. Do not use stairs or walkways as storage space.
- Do not allow water or ice to accumulate on these surfaces to prevent users from slipping and falling.
- Ensure that steel structures are adequately painted or pre-conditioned to prevent corrosion and subsequent reduction in load carrying capacity.

- Consider coating concrete or steel walking surfaces with an anti-slip finish.
- Maintain all doors or gates to guarantee proper operating condition. Also, do not lock doors or gates that are necessary for emergency exit.
- Ensure that all metal components are isolated from any electrification system.

3.   Ventilation Structures and Emergency Egress Shafts

In rail transit systems, ventilation structures are often utilized in urban settings where tunnel intake or exhaust occurs through grates in the sidewalk or street above the tunnel. Subsequently, the air travels though a shaft structure that extends to the depth of the tunnel, which could be constructed from any of the same materials available for tunnel construction. If the routine inspection identifies structural deficiencies in this shaft or in emergency egress shafts, then they can be repaired using techniques given in Chapter 4. Otherwise, a preventive maintenance program should be instituted to ensure the following:

- Keep areas clean and free of debris. Do not use as storage spaces.
- Maintain grates at top of vent structures to prevent corrosion and dislodging, which could pose a safety hazard to pedestrians walking across the grating.
- Perform maintenance on any fans at top of vent structures as described in Section B of this chapter.

CHAPTER 4:
REHABILITATION OF STRUCTURAL ELEMENTS

This chapter describes various methods for repairing specific deficiencies in structural elements within a tunnel. Water infiltration is the most common cause of deterioration. However, deficiencies could be the result of substandard design or construction, or the result of unforeseen or changing geologic conditions in the ground that supports the tunnel. Another common reason for repairs is the fact that many tunnels have outlived their designed life expectancy and therefore the construction materials themselves are degrading. Due to the fact that there are different causes for the degradation, the method of repair could vary.

This chapter provides in-depth discussions and recommendations for repairing tunnels that are being deteriorated by water infiltration (Section A). In addition, a detailed explanation of the different types of concrete deficiencies and methods for their repair is provided in Section B. Section C addresses the issue of repair for specific types of liner construction.

A. WATER INFILTRATION

1. Problem

Since many tunnels are constructed deep in the ground and often below the groundwater table, controlling water infiltration is of great concern to tunnel owners. Consequently, water infiltration is the underlying cause of most deterioration of the tunnel structure and components. Water infiltration can occur in all types of tunnel construction. Even tunnels that are designed to be waterproof, such as immersed tube tunnels that are placed in a trench at the bottom of a body of water, can develop leaks due to inadequate connection/joint design, substandard construction, and deterioration of the waterproof lining due to chemical or biological agents in the water or from tears caused by tunnel settlement. Most tunnels are designed with the foreknowledge that water will exist in the ground, but it is prevented from entering the tunnel by providing drainage mechanisms around the exterior of the lining or embedded within the joints. As ground water flow patterns change over time and drains become clogged with sediment, the water is bound to find its way into the tunnel through joints or structural cracks.

Another scenario that may occur in a few urban settings is that the elevation of the ground water table may rise due to the accumulating effects of basements of surrounding buildings being made relatively waterproof and the city's water supply needs being met by reservoirs many km (miles) away instead of through groundwater extraction. This could cause a tunnel that was designed to be above the water table to experience hydrostatic forces that it is unable to resist and subsequently water infiltration becomes a problem.

It should be noted that in the 1960s some tunnel owners began to develop maximum allowable rates of water infiltration to be used as a guide to determine original design and subsequent repairs if the amount of infiltration increases. One such owner was the Bay Area Rapid Transit (BART) system in California; they set a limit of 0.8 liters/minute per 75 linear meters (0.2 gpm per 250 linear feet) of tunnel. This translates to 3 liters/minute per 300 linear meters (0.8 gpm per 1000 linear feet) of tunnel. Other tunnel owners have adopted this criteria while still others may use a limit of 3.8 liters/ minute per 300 linear meters (1 gpm per 1000 linear feet) of tunnel. These limits are for reference purposes only, with the main emphasis for determining repair needs placed on the location of the leak and the condition of the tunnel components that are affected.

2. <u>Consequences of Water Infiltration</u>

As can be expected, nothing positive occurs when water infiltrates into a tunnel. The negative consequences can vary from minor surface corrosion of tunnel appurtenances to major deterioration of the structure and thus decreased load carrying capacity of the tunnel. Most tunnels have problems that fall somewhere in between. Below is a list of possible forms of tunnel degradation or safety risks that can result from water infiltration.

- Cement and sometimes aggregates of concrete liners are eroded causing the structure to be weakened.
- Reinforcement steel with poor or inadequate cover corrodes and causes delamination and spalling of the concrete cover.
- Bolts that connect segmental linings can corrode and fail.
- Masonry units and mortar can be very susceptible to water deterioration and can swell or become brittle depending on chemicals in water.
- Steel segmental liners or steel plates can experience section loss if exposed to both moisture and air.
- Fine soil particles can be carried through cracks with the water, creating voids behind the liner, which can cause settlement of surrounding structures and/or cause eccentric loading on tunnel that can lead to unforeseen stresses. These fine particles can also clog drains in or behind the lining.
- Fasteners of interior finishes or other appurtenances (fans, lights, etc.) can corrode and pose danger to a motorists or trains traveling through the tunnel.
- Water may freeze on roadway and safety walks or form icicles from the tunnel crown, all of which endanger tunnel users (Figure 4.1).
- Frozen drains can cause ground water to find or create a new location to enter the tunnel, which may be undesirable.
- Road salts carried by vehicles into highway tunnels, along with the presence of infiltrated water, can increase deterioration of the structure, especially the invert.
- Rate of corrosion for tunnel components of rail transit tunnels can be increased by the presence of stray current from electrified traction power systems.

Figure 4.1 – Ice formation at location of water infiltration in plenum area above ceiling slab.

3.   Remediation Methods

In general, there are three options that a tunnel owner must consider for remediation of a water infiltration problem. The three alternatives are: short term repairs, long term repairs or, as a last resort, reconstruct all or portions of the tunnel lining that is causing the problem using methods of waterproofing that incorporate newer technologies. It should be noted that the alternative classifications are given for descriptive purposes and that overlaps between them do exist.

Since the first and second alternatives are the most common and usually most cost effective, a more detailed development of the current methods and some associated details will be given. The third option will not be discussed in as much detail as the first two but will include a brief discussion of some current technologies being used on new tunnel construction. To determine the most cost efficient method of repair for a particular situation, a specific cost analysis should be performed that considers the costs over the life of the tunnel. For guidance in this effort, a brief explanation of life-cycle-cost methodology is given in Appendix A.

a)   Short Term Repairs

For certain situations, it might be necessary to redirect infiltrated water to the tunnel drainage system on a temporary basis until further investigation can be performed and a more long term solution implemented. It should be noted that certain tunnels, whether due to deficiencies in design or construction or a change in the ground water table, will not be able to stop the water infiltration completely without a total restoration or reconstruction of the tunnel lining or at least significant portions where water infiltration is a problem. Therefore, some tunnels may have to rely on a long term system that conveys the water rather than prevents the water from entering the tunnel. Long term systems will be discussed more in-depth in the next section of this chapter, but the following paragraphs cover a few methods for temporarily diverting the infiltrated water.

(1)   Drainage Troughs

If leaks are occurring in joints at the tunnel crown in a direction perpendicular to the tunnel length, then neoprene rubber sheets can be attached to the tunnel lining with aluminum channels. The sheets can be directed to channel the water to the side of the tunnel where it can flow into the tunnel drainage system (Figure 4.2). A similar method utilizing metal drainage troughs is sometimes used to redirect isolated areas of infiltration to the drainage system.

Figure 4.2 – Temporary drainage systems comprised of neoprene rubber troughs and 25 mm (1 in) aluminum channels.

(2)   Plastic Pipe Network

Another rather rudimentary method is to use plastic piping with one end inserted into the concrete at the main concentration of the leak. The piping can be hooked together in a network that conveys the water to the primary drainage system (Figure 4.3).

Pipe inserted into concrete

Conveyance pipe to primary drainage system

Figure 4.3 – Temporary drainage system comprised of 50 mm (2 in) plastic pipe.

b) <u>Long Term Repairs</u>

Since water infiltration is an ongoing problem for tunnel owners, there have been a wide variety of methods and materials used to prevent the water from entering the tunnel and causing undesirable degradation. Multiple techniques have not performed favorably over the long term, but that does not necessarily mean that the method utilized was the problem. Many different factors are involved in determining which method should be used that are site specific in that the cause and volume of the water infiltration will help determine how to properly prevent it. One method might work very well for one tunnel but not another. Therefore, it is suggested that a detailed study be performed on major leaks to determine the source and amount of water leakage, and the cause and exact location of the leak. This, along with knowing the type and condition of the materials that make up the tunnel lining structure, will help determine how to address the problem. Also, the method of preparing the surface and the procedure for installing the waterproofing system should be investigated to help determine which system should be used. The following paragraphs describe a few methods that have been used to address water infiltration problems for the long term.

(1) <u>Insulated Panels</u>

Insulated panels have been successfully used to line exposed rock tunnels to allow the water to flow behind the insulation down to the primary drainage system, while being insulated to prevent water from freezing. An example of this installation is a tunnel in the Pennsylvania Mountains that used two-inch-thick, 2.4 m by 9.6 m (8 ft by 32 ft) panels of Ethafoam insulation that was secured to the rock using 12 mm (½ in) diameter galvanized steel pins set into the rock on a .9 m (3 ft) square grid (Figure 4.4). It should be noted that use of this type of system would reduce the interior clearances within the tunnel.

Figure 4.4 – Insulated panels used as a waterproofing lining to keep infiltrating water from freezing. (Photo courtesy of Tunnels & Tunnelling International)

(2)   Waterproofing Membrane

As an addition to the method given above, a continuous, flexible membrane can be used as the waterproofing layer that allows the water to flow towards the main tunnel drainage system. The specific process that has been effectively used involves placing a geotextile material against the existing tunnel interior, then a PVC waterproofing membrane, followed by a layer of material that will protect the membrane, such as shotcrete or other fire-retardant and protective materials. The term geotextile stands for a wide variety of materials which are normally synthetic and whose main purpose are to provide a drainage gallery outside the waterproofing membrane through which the infiltrating water can freely pass. The geotextile layer also provides a physical protection of the waterproofing membrane. Refer to Figure 4.5 for a detail of this system.

This system requires a relatively smooth surface to attach the membrane to, without projections that could potentially puncture the membrane. It is suggested that mock-up trials be performed to ensure that the components of the system achieve adequate bond to each other, especially the application of a protective layer on the inside of the membrane. If shotcrete is used a minimum membrane thickness might be required as well as limiting the aggregate size in the shotcrete. If a fire retardant protective material is applied in sheets then the connection of this material through the membrane must be properly sealed to prevent water infiltration through this joint.

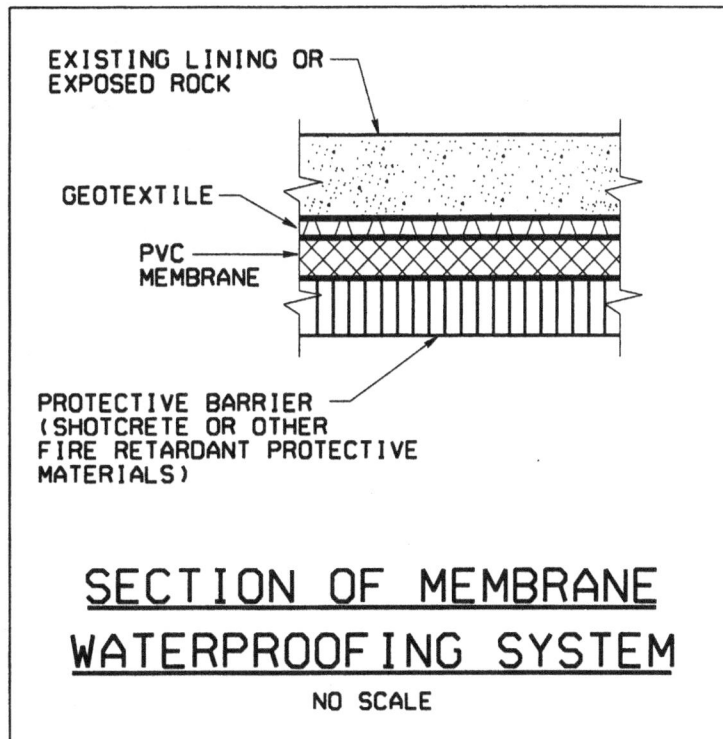

EXISTING LINING OR
EXPOSED ROCK

GEOTEXTILE

PVC
MEMBRANE

PROTECTIVE BARRIER
(SHOTCRETE OR OTHER
FIRE RETARDANT PROTECTIVE
MATERIALS)

SECTION OF MEMBRANE WATERPROOFING SYSTEM

NO SCALE

Figure 4.5

This system can also be supplemented by inserting pressure relief holes into the surrounding soil/rock that provide a path of least resistance for the infiltrating water, so that adverse hydraulic pressures are not allowed to build up behind the liner. Additionally, a temperature controlled heat strip can be attached to exposed drainage pipes that prevents freezing of water in pipes and subsequent back up of water.

It should be noted that material types other than those stated have been used successfully, which include both preformed sheet materials and liquid applied materials for the waterproofing membrane layer. Therefore, research of current material technology should be performed prior to selecting the individual components of the waterproofing membrane system. The system chosen may need to be site specific given the possible presence of hydrocarbons or other chemicals that could adversely affect the membrane material. Some of the other materials available include polyolefin, which includes polyethylene and polypropylene, and sprayable polymer membranes. The manufacturer of the materials should be consulted and they should be able to supply material specifications and case histories of where the material might have been used successfully. Committee 515 of the American Concrete Institute (ACI) has also developed a guide for the use of waterproofing membrane systems. This is recommended as an additional source of information; however, it does not specifically describe tunnel applications.

The success of this system is primarily dependent on the ability to install a continuous membrane and whether a proper connection of this membrane to the tunnel drainage system is achieved. The membrane chosen must also be able to withstand any future movement of the structure without reflective cracking and must be resistant to chemical or biological attack from the infiltrating ground water.

(3)   Crack/Joint Injection

The most common method for preventing water infiltration in concrete linings is to inject the crack/joint with a particle or chemical grout. Particle grouts are very fine cementitious grouts that produce nonflexible fillers that prevent water from penetrating the crack/joint. Since these grouts are nonflexible, they are not recommended for any location that might experience structural movements in the future. Chemical grouts on the other hand can be highly flexible and also have low viscosities that enable them to be injected into very thin cracks. Chemical grouts are expensive, sometimes toxic or flammable and require a high degree of skill for proper application; therefore, an understanding of the chemical properties and their suitability for the desired application is essential.

Even with the drawbacks of some chemical grouts, their performance in stopping water infiltration is significantly superior to particle grouts; therefore, they are used more frequently. It is important to note that if chemical grouts are allowed to dry out they may not be as effective. This could happen if the source of the water infiltration is diverted or the ground water elevation drops below the crack location. In the event of a dry crack, repair methods discussed in Section B of this chapter should be considered.

Of the chemical grouts developed to date, the polyurethane, reactive grouts have performed the best for tunnel applications. This type of grout expands into a foam at the presence of water and subsequently seals off the crack, not allowing water to pass through. This foam is also moderately resistant to tensile forces; therefore it can expand when and if a crack/joint continues to open further. Figure 4.6 shows and explains the procedure for properly injecting a vertical or overhead crack/joint with a chemical grout. It has been found that when applying pressure to inject the grout that low pressure for an extended period is better than high pressure for a short period. The latter can result in further damage to the concrete.

In addition to polyurethane chemical grouts, acrylate esters are also being used to inject cracks. The esters have an advantage over the polyurethanes in that they form a gel upon reaction with the water and serve as a barrier to water penetrating a crack. The esters will also not dry out as can occur with polyurethane grouts as described earlier. For this reason, a site specific investigation will need to be conducted to determine which material is most cost-effective over the long term.

It should be noted that cracks in masonry liners can also be injected, but often times other methods of repair are more effective for masonry over the long term. These methods will be discussed in Section C of this chapter.

NOTES

1. REMOVE LOOSE MATERIAL FROM CRACK.
2. DRILL 15 MM (5/8 IN) DIA. INJECTION PORTS AT 45 DEG. ANGLE TO CRACK, ALTERNATING SIDES.
3. INSTALL MECHANICAL PACKER IN INJECTION PORT.
4. SEAL SURFACE OF CRACK WITH LOW-MODULUS GEL IF CRACK IS ACTIVELY LEAKING.
5. FLUSH CRACK WITH CLEAN WATER.
6. INJECT LIQUID URETHANE OR ACRYLATE ESTER RESIN INTO LOWEST MECHANICAL PACKER WITH HAND OPERATED HYDRAULIC PUMP UNTIL GROUT CAN BE SEEN AT THE NEXT INJECTION PORT UP.
7. REPEAT PROCESS UNTIL ENTIRE CRACK IS INJECTED.
8. REGROUTING MAY BE PERFORMED FOR UP TO A WEEK AFTER INITIAL GROUTING.

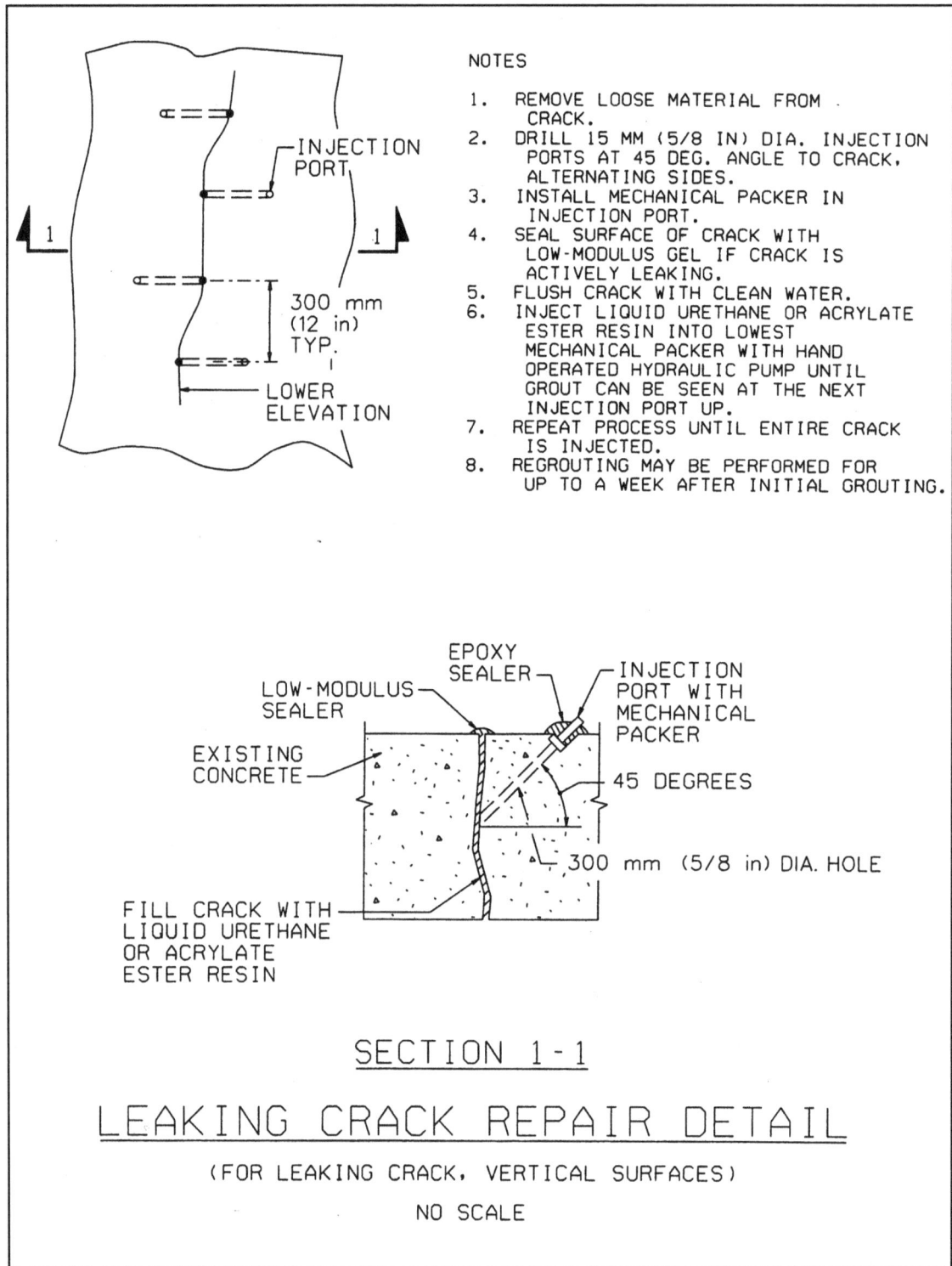

INJECTION PORT

300 mm (12 in) TYP.

LOWER ELEVATION

LOW-MODULUS SEALER

EPOXY SEALER

INJECTION PORT WITH MECHANICAL PACKER

EXISTING CONCRETE

45 DEGREES

300 mm (5/8 in) DIA. HOLE

FILL CRACK WITH LIQUID URETHANE OR ACRYLATE ESTER RESIN

SECTION 1-1

LEAKING CRACK REPAIR DETAIL

(FOR LEAKING CRACK, VERTICAL SURFACES)

NO SCALE

Figure 4.6

(4)   Soil/Rock Grouting (Back-Wall Grouting)

As an alternative to injecting a crack/joint (which is generally successful for stopping the leak through the injected crack/joint, but can force the water along the path of least resistance towards another crack/joint), similar materials can be injected through the liner into the soil/rock beyond. The goal of this method is to provide a protective barrier on the outside of the tunnel lining either in specific crack/joint locations or over an entire segment of the tunnel. The material that is injected can form this protective barrier or the injected material can introduce cohesion into the soil, which makes the soil itself impermeable.

The procedure for this method consists of drilling holes perpendicular to and through the liner on a predetermined pattern (based on ground conditions and amount of water present), and installing mechanical injection packers. Then, a grout is injected into the soil/rock and maintained at a constant pressure for a prescribed amount of time to allow the grout to penetrate small cracks in the soil/rock. There are different grouts that are available and a site-specific investigation is necessary to determine which one is best suited for the particular conditions. Some of the available grouts are:

- Microfine cement grouts
- Polyurethane chemical grouts
- Acrylate ester resin chemical grouts
- Acrylamide-based chemical grouts (highly toxic).

Typically the chemical grouts are more expensive; therefore, the cement grouts can be used for areas where voids exist behind the liner and large volumes of grout are required.

In the case of a steel or cast iron liner, the existing grout plug holes should be used as the location for the new grout placement, since the liner would not have been designed to handle additional holes being drilled through it.

One example of this system would be the Bay Parkway Bridge in New York City which is 45 m (150 ft) wide and has soil cover over a rail line running underneath that essentially forms a tunnel. The New York City Department of Transportation chose to utilize the acrylate ester resin chemical grout as the injection material and they used a .6 m to .75 m (2 ft to 2-½ ft) center-to-center spacing for their injection pattern. To date, this repair has performed well and other similar applications are being considered. Another example of this system being used successfully would be in the subway tunnels of the Toronto Transit Commission

(TTC). For their situation they chose to use an acrylamide based chemical grout and they had specialty grouting work cars fabricated to condense and mobilize the operation for brief nighttime work periods when the tunnel could be closed. Their experience has shown that this can be a reliable method of stopping water infiltration.

It should be noted that this system could be used in conjunction with other systems. An example would be to back-wall grout a particular area and therefore force water to flow to a predetermined point where a drainage system could be installed. More details for installing drains within the liner are given in the next method.

(5)   Crack/Joint Repair

If water infiltration through cracks/joints in concrete linings cannot be stopped by injecting the crack/joint as described previously because of excessive movement which surpasses the tensile strength of the grout material used, then another approach is to convert a crack into a joint that allows differential movement of the concrete, and add waterproofing components to the existing joints. Figure 4.7 portrays a method of routing out the crack or joint to a specific depth and then properly sealing off the water infiltration with successive layers of different impervious materials. The finished product will look and behave like a joint in that it will allow for some differential movement and will be watertight. As with the other repair techniques, a registered professional engineer should review and approve the application of this method to the specific site location. This is especially true for this method due to the possible weakening of the structural capacity of the lining depending on where and what direction the crack is located.

Figures 4.8 and 4.9 deal specifically with cracks and joints respectively and begin by routing or cleaning in the case of a joint. The difference with this method is the addition of a semi-perforated pipe that is inserted into the crack/joint, which enables the infiltrating water to be collected from the exterior side of the pipe and exported into the tunnel drainage system at the bottom of the crack. The pipe can be covered with a neoprene rubber sheet (liquid neoprene is also applicable) on the exterior of the concrete or mastic and impervious mortar can be used to make the repair look just like a normal joint.

(6)   Segmental Joint Repair

Segmental liners can be made of either precast concrete, steel, or in the case of older tunnels–cast iron. Water infiltration generally occurs at the joint location where the original lead, mastic, or rubber seal has failed.

This can be corrected by repacking the joint with new sealing material and installing new gaskets at boltholes. Cracks and joints can also be injected with particle or chemical grouts as discussed previously. In the case of precast concrete segments, the cracks are injected similar to method (3). In addition, for single-pass liner systems with any of the three segmental liner types, the processes described in method (4) can be implemented on the exterior of the liner with the precautions noted.

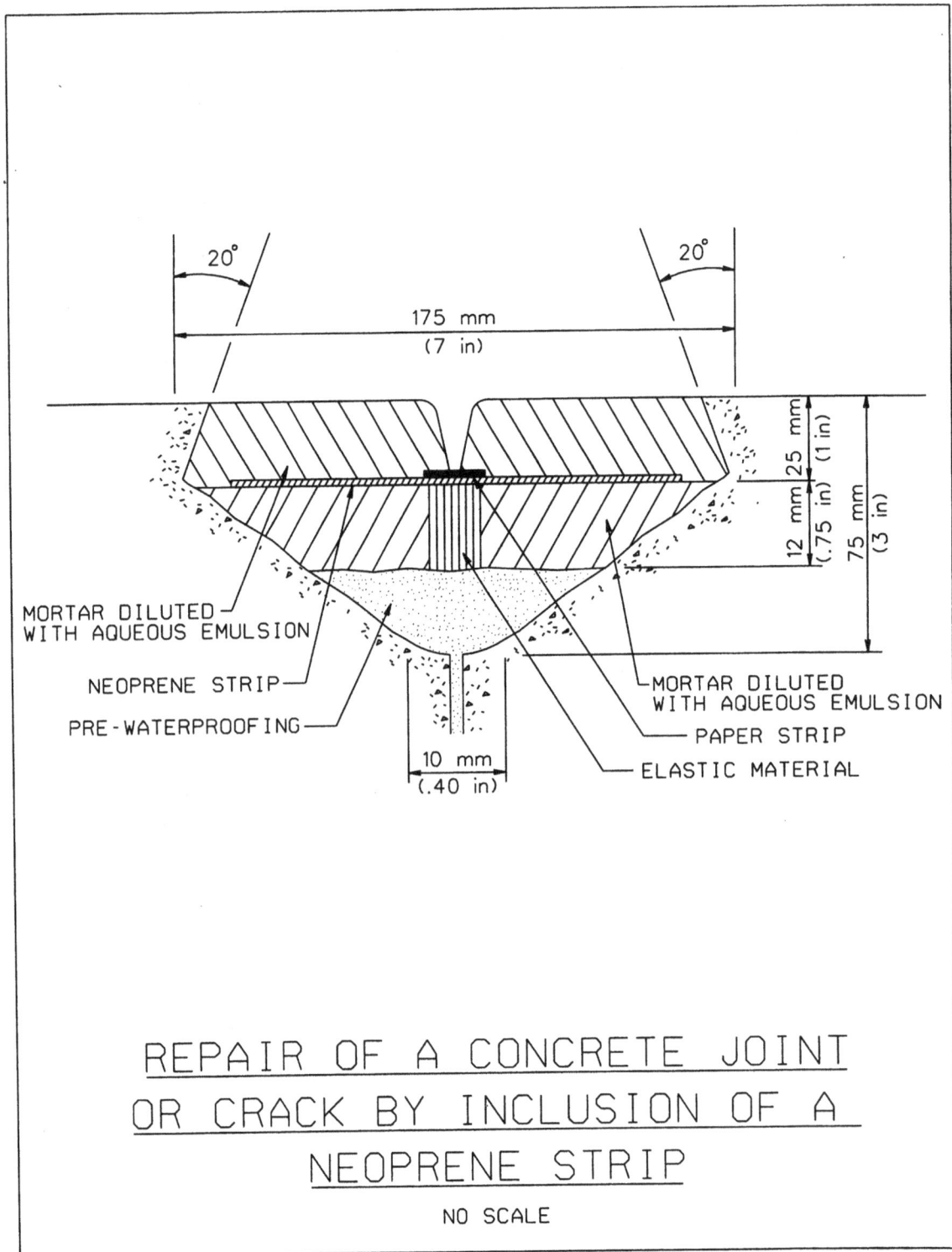

REPAIR OF A CONCRETE JOINT OR CRACK BY INCLUSION OF A NEOPRENE STRIP

NO SCALE

Figure 4.7

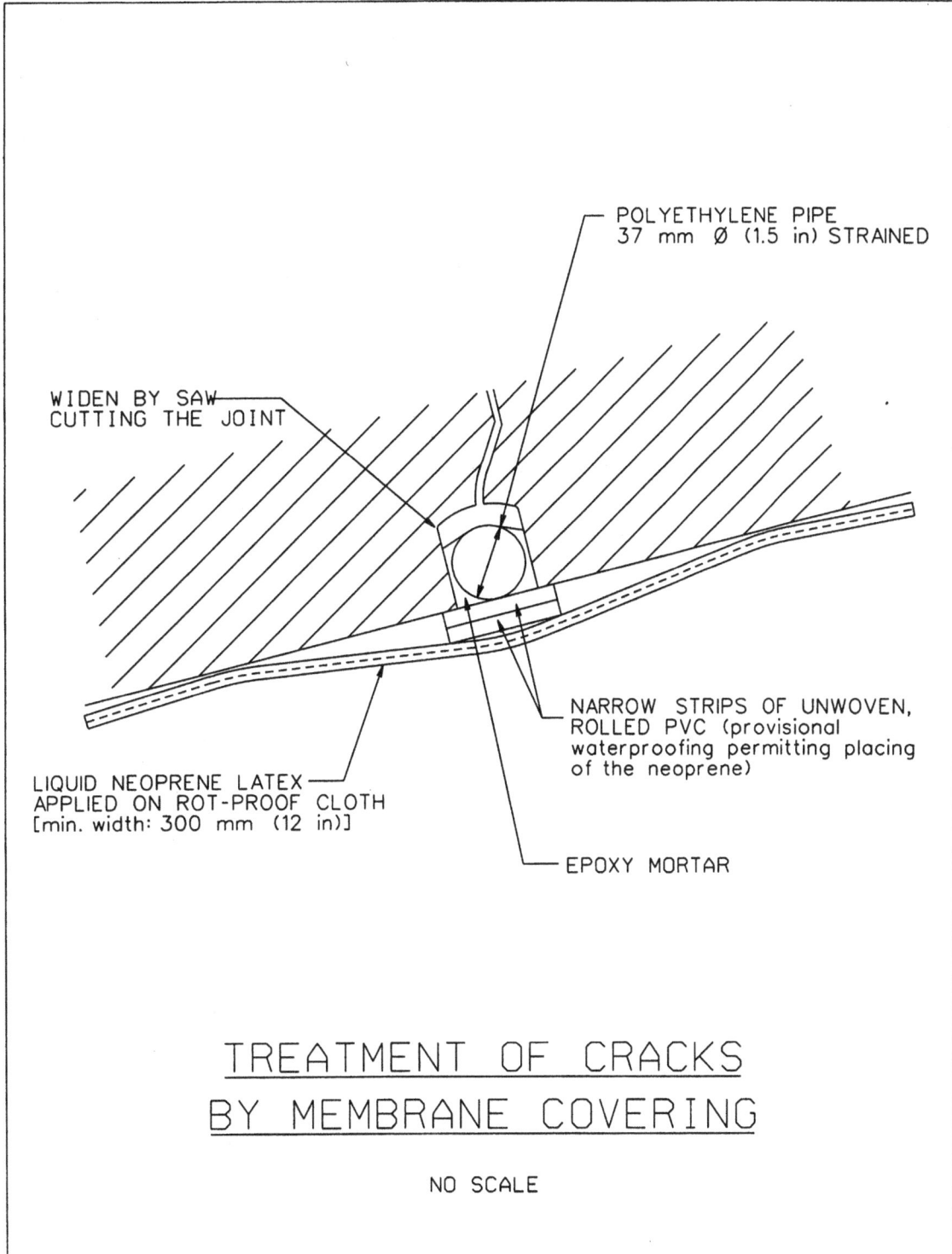

POLYETHYLENE PIPE
37 mm Ø (1.5 in) STRAINED

WIDEN BY SAW
CUTTING THE JOINT

NARROW STRIPS OF UNWOVEN,
ROLLED PVC (provisional
waterproofing permitting placing
of the neoprene)

LIQUID NEOPRENE LATEX
APPLIED ON ROT-PROOF CLOTH
[min. width: 300 mm (12 in)]

EPOXY MORTAR

TREATMENT OF CRACKS
BY MEMBRANE COVERING

NO SCALE

Figure 4.8

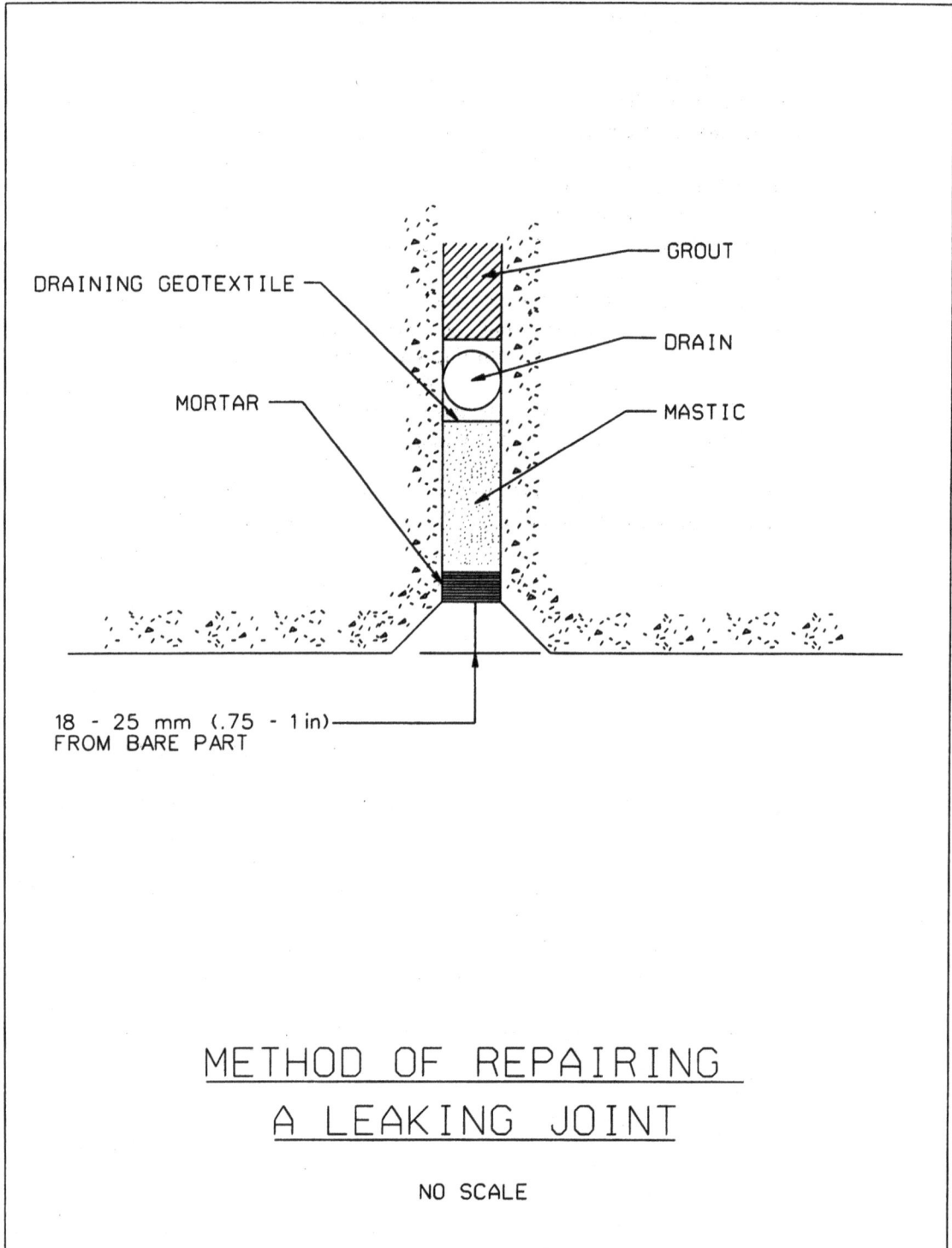

DRAINING GEOTEXTILE ―

GROUT

DRAIN

MORTAR ―

MASTIC

18 - 25 mm (.75 - 1 in)
FROM BARE PART

METHOD OF REPAIRING
A LEAKING JOINT

NO SCALE

Figure 4.9

c)   Reconstruction and New Construction

If the tunnel degradation has advanced to a point where repairing numerous localized areas of the liner becomes cost prohibitive, it may be necessary to reconstruct larger areas using different techniques. This could include shotcrete or pumping plasticized concrete within a form liner. There are several relatively new technologies that are being used for new tunnel construction that can also be incorporated into reconstruction procedures, with some modifications. These methods generally attempt to prohibit the water from infiltrating the final liner and thus entering into the tunnel space. This is accomplished by collecting the water and draining it away either within the liner or on the exterior of the tunnel. The latter method is less common because the drains can become clogged with fine soil particles. In addition, using an exterior drainage system in a tunnel below the ground water elevation is normally not effective over the long term because of the ability for water to penetrate very small cracks that develop between drains.

There are various detailed techniques that will only be explained briefly, although many of these are complex in nature. Furthermore, should an extensive repair be needed, it is recommended that a specialized consultant be obtained to develop possible solutions that are specific to the tunnel in question. The following paragraphs describe available systems for extensive lining reconstruction or that are also applicable for new tunnel construction.

(1)   Shotcrete Applications

The use of shotcrete in tunnel construction has greatly increased since the advent of the Sequential Excavation Method (SEM) and the improvement of the shotcrete materials and application processes used. A few of the general material classifications for shotcrete are–cementitious, latex/acrylic-modified, or two-component epoxy. Shotcrete can also be used in tunnel rehabilitation in various forms. One method is to simply coat the entire interior of the tunnel walls and ceiling with a mix design that makes the cured shotcrete relatively impervious to water. This method has some drawbacks that include decreasing the tunnel clearances and trapping the moisture inside the original liner. Trapped moisture can lead to deterioration due to chemical reactions between the water and the liner material, especially in masonry.

Another more in-depth procedure is to remove all or portions of the existing liner, replace it with a structural layer of shotcrete, then place a geotextile layer and waterproofing membrane (either sheet membrane or sprayable polymer membrane), and finally provide a protective, non-structural finish liner of shotcrete on the inside that initially adheres to the waterproofing membrane during curing. As mentioned previously, the membrane thickness and shotcrete aggregate size may have restrictions placed on them in order to ensure that the membrane is not damaged during the shotcreting procedure. It is possible to place another geotextile layer or other protective material on the inside of the membrane, but attachment of this layer is difficult since the attachment mechanism has to puncture the membrane The thickness of this liner is dependent on the tunnel size and shape and the amount of water infiltration that is expected. It is recommended that a detailed site investigation be performed to determine if this final lining will need to resist any hydrostatic loadings. This method allows water that penetrates the initial liner to be directed down the tunnel along the waterproofing membrane to the primary tunnel drainage system. The existing liner can be removed with traditional demolition techniques or, depending on the depth of removal desired, a modern laser-controlled cutterhead mounted on a boom as shown in Figure 4.10 can be used to remove precise depths of masonry, concrete or rock.

Figure 4.10 – Laser controlled cutter for removing portions of existing tunnel liner. (Photo courtesy of Tunnels & Tunnelling North America)

(2)   Joint Control

Deteriorated joints can be repaired as described previously in Chapter 4, Section A, Part 3b(5). It is not often that there is an opportunity to completely reconstruct a joint in an existing tunnel. However, when there is a complete tunnel reconstruction or new tunnel construction, the joints can be fitted with a new system that allows the joint to be initially injected with chemical or particle grouts and to be reinjected at any future time that the joint might begin to leak due to settlement of the structure. Also, products exist that can be inserted at anticipated crack locations that actually facilitate crack development at that location. Once the crack occurs, the product can be injected with a chemical or particle grout to stop water infiltration.

(3)   <u>Concrete Design</u>

One of the most effective methods of preventing water infiltration in reconstruction or new construction is to properly design the concrete or shotcrete mix to approach impermeability and to not be as susceptible to cracking. This is primarily done by ensuring adequate reinforcement and limiting the water/cement ratio to 0.45. Other considerations include the use of water reducing and shrinkage reducing admixtures. Another admixture that is increasing in usage is a waterproofing additive. This admixture reacts with the fresh concrete to produce crystalline formations throughout the cured concrete that resist the penetration of water.

When major repairs or reconstruction is required, a detailed site-specific investigation should be undertaken to determine what methods and materials can be applied based on current research and experience.

B.   <u>CONCRETE REPAIRS</u>

As concrete deteriorates, it is important that proper repairs be made to avoid further degradation of the structure. The repairs must be durable, easy to install, capable of being performed quickly during non-operating hours, and cost-effective. The repairs included in this manual are commonly used and have been performed in various tunnel locations.

The defect must first be evaluated to determine the cause and the severity of the deterioration, in order to select the best repair method. Factors affecting the repair are the severity to which the concrete has deteriorated, whether water infiltration is the cause, location of the repair to be completed, and the structural impact of the defect. Repairs should not be made until the cause of the defect has been determined and the situation remedied, or the same problem may repeat itself in the newly repaired concrete.

It should be noted that many concrete linings in highway tunnels have an additional tunnel finish that covers the concrete and therefore may hide the extent of the deterioration. This finish commonly is porcelain tile or prefabricated metal or concrete panels. Therefore, a repair analysis will need to account for the replacement or repair of the finish as well. Concrete deterioration in tunnels may be caused by any of the various factors listed below.

·   <u>Water Infiltration</u> – Refer to Chapter 4, Section A for a discussion of the negative effects of water infiltration and suggested methods of repair.

·   <u>Corrosion From Embedded Metal</u> – Several factors contribute to accelerate the corrosion of embedded steel, such as oxygen, water, stray electrical currents, chemicals, chlorides, and low pH (acidity). Once the corrosion has begun, signs of this problem are delamination (a separation of the concrete from the embedded·   steel), surface spalling, or cracking. Cracks may have existed previously that permit deleterious elements access to the reinforcement steel.

- Disintegration of Material – Certain chemicals like acids, alkaline solutions, and salt solutions are common enemies of concrete. Acid attacks concrete by reacting with the calcium hydroxide of the hydrated Portland cement. This reaction produces a water-soluble calcium compound, which is then leached away.

  Porous concrete will absorb water into small capillaries and pores. Once there, the water freezes, then expands and exerts tension forces on the concrete. Near the surface small flakes of concrete will break away causing further exposure and eventual spalling and removal of aggregate with the process continuing inward.

- Thermal Effects – Thermal loads cause the concrete to expand and contract putting undue stress on the concrete. This expansion and contraction can lead to cracking. However, due to the relatively uniform environment within a tunnel, this form of degradation of the concrete is limited to areas near portals and possibly within air plenums where temperature fluctuations are more likely.

- Loading Conditions – Load placement will have varying effects on concrete. For continuous concrete spans in roadway slabs over air ducts, cracks may develop over the underlying steel support on the slab topside. In the center of the span, cracks will develop on the underside of the slab. Shear cracks may also develop near the support.

- Poor Workmanship – Workmanship is critical to overall concrete performance. If the reinforcement steel is placed improperly, if there is insufficient vibration to consolidate the concrete, if the concrete is permitted to segregate when placing, or if the concrete is not finished or cured properly, then the strength and long-term durability of the concrete will be affected.

Once the defect has been evaluated and the cause determined, one of the following potential repairs should be implemented:

1. Crack

The most common defect found in concrete is a crack. For cracks where water infiltration or moisture is present, see Chapter 4 Section A for methods of repair. For cracks that are void of water, and movements are not expected, the crack can be filled with an epoxy resin. For cracks on a horizontal surface, the crack may be gravity filled with epoxy by constructing a temporary dam (see Figure 4.11). However, the underside of the concrete surface may need to be sealed, if it is accessible, to prevent the resin from running completely through the crack. For vertical and overhead cracks, a paste gel is placed on the surface of the crack, around the injection ports to contain the resin that fills the crack. See Figures 4.11 and 4.12 for examples of this repair.

PROVIDE TEMPORARY DAM EACH SIDE OF
CRACK TO CONTROL FLOW OF GRAVITY.
FEED EPOXY RESIN INTO CRACK.

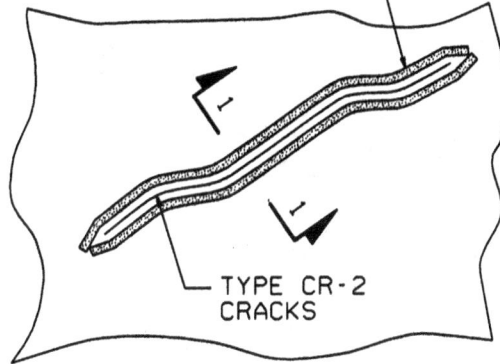

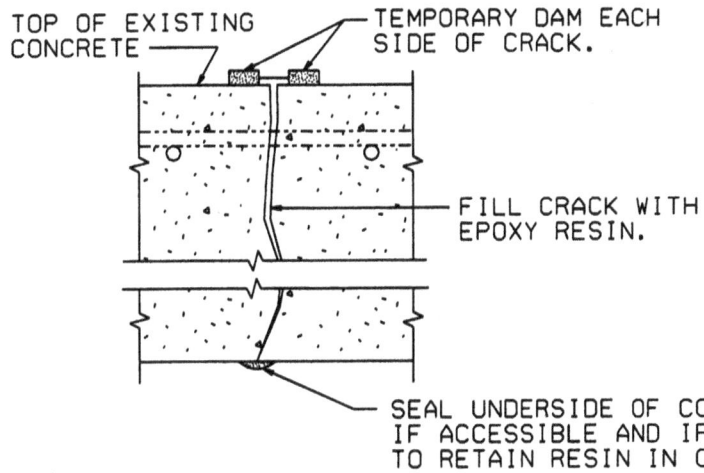

TYPE CR-2
CRACKS

TOP OF EXISTING
CONCRETE

TEMPORARY DAM EACH
SIDE OF CRACK.

FILL CRACK WITH
EPOXY RESIN.

SEAL UNDERSIDE OF CONCRETE
IF ACCESSIBLE AND IF NEEDED
TO RETAIN RESIN IN CRACK.

SECTION 1-1

HORIZONTAL SURFACE
CRACK REPAIR DETAIL

(FOR CRACKS 0.8 MM (1/32 IN) WIDE AND GREATER)

NO SCALE

Figure 4.11

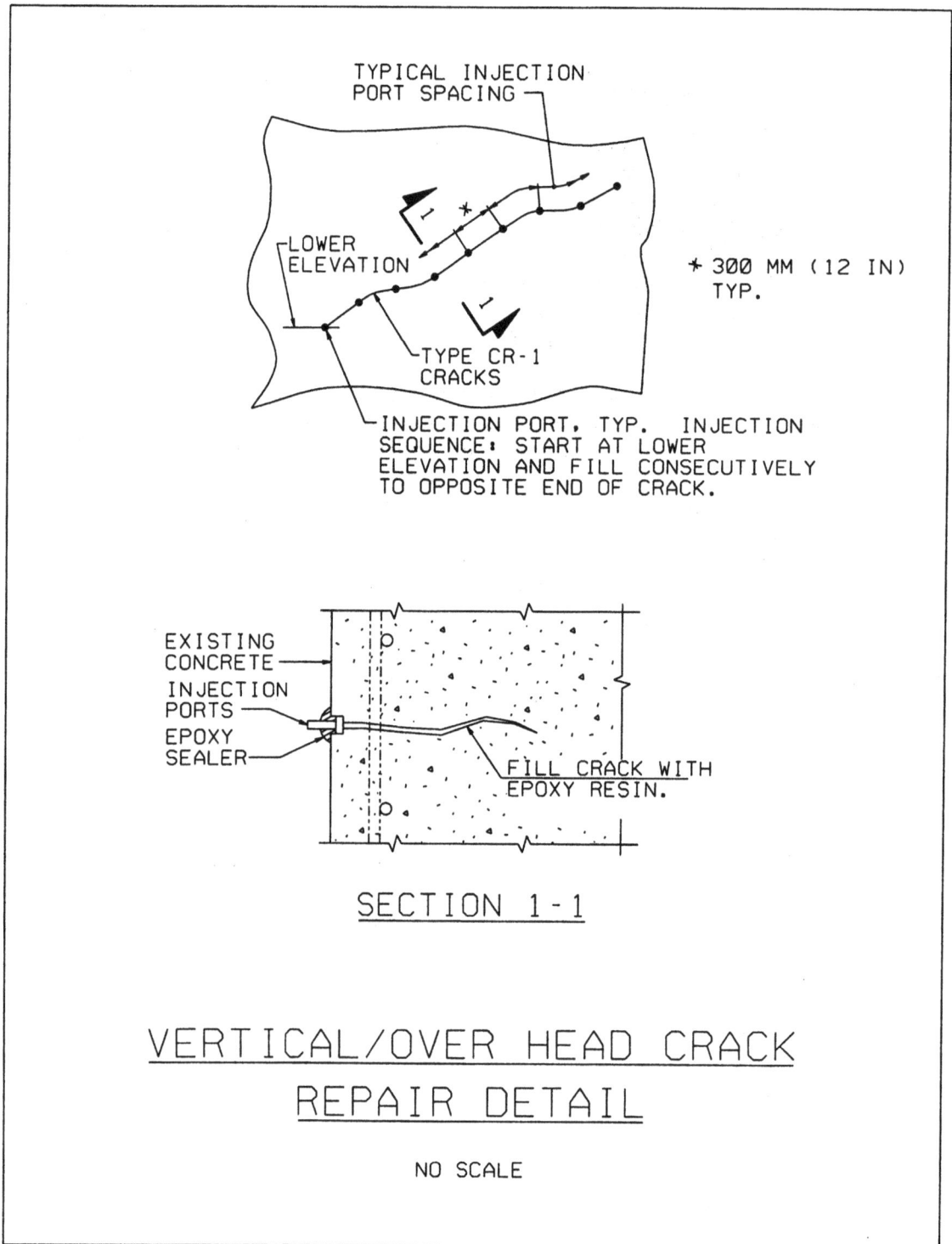

TYPICAL INJECTION
PORT SPACING

LOWER
ELEVATION

TYPE CR-1
CRACKS

✳ 300 MM (12 IN)
TYP.

INJECTION PORT, TYP.   INJECTION
SEQUENCE: START AT LOWER
ELEVATION AND FILL CONSECUTIVELY
TO OPPOSITE END OF CRACK.

EXISTING
CONCRETE

INJECTION
PORTS

EPOXY
SEALER

FILL CRACK WITH
EPOXY RESIN.

SECTION 1-1

VERTICAL/OVER HEAD CRACK REPAIR DETAIL

NO SCALE

Figure 4.12

2.   Spall

A spall is an irregular shaped depression in the concrete in which the fracture is parallel, or slightly inclined, to the surface. It is caused by the separation and removal of a portion of the surface concrete, typically due to corroded reinforcement steel, where the tensile stresses in the concrete exceed the tensile capacity. However, some spalls may occur that do not have any exposed steel. Depths of spalls vary and for repair purposes can be classified as either shallow or deep. A shallow spall typically penetrates less than 50 mm (2 in) into the concrete, whereas deep spalls penetrate 50 mm (2 in) or more into the concrete and usually expose the reinforcement steel within. Reinforcement steel can also be exposed in a shallow spall if it was originally placed too close to the surface of the concrete, resulting in a pop off of the concrete cover.

Special attention needs to be given to determining the cause of any corrosion on the reinforcement steel. If corrosion is due to water infiltration from the exterior of the tunnel, then the methods and materials given in this section may not be adequate to resist the effects of future infiltration. For this situation, it is necessary to address the water infiltration using methods given in Chapter 4, Section A. But, if a complete restoration of the original concrete surface is desired, the following methods can be used.

If the inspector recommends that the spalls should be repaired to preserve the integrity of the concrete, the following procedures may be utilized:

   a)   Shallow Spall With No Reinforcement Steel Exposed (See Figure 4.13)

        This repair is typically performed for aesthetic reasons and not necessarily for structural integrity of the lining. Suggested steps include:

   - Remove all loose or delaminated concrete on the spall surface.
   - Clean the concrete surface of deleterious materials.
   - Sawcut around the spalled area on a 20-degree angle.
   - Place polymer repair mortar in the spall to original concrete depth.

   b)   Shallow Spall With Reinforcement Steel Exposed (See Figure 4.14)

        If the exposed reinforcement steel is only slightly corroded with no significant section loss, then this repair method can be used. If, however, the corrosion appears to be deeper than the current spall depth, or if the spall extends behind the reinforcement steel, it is recommended that the extent of the corrosion be determined and the spall be repaired by the method given in Part c). Suggested repair steps include:

   - Remove all loose or delaminated concrete around the exposed reinforcement steel.
   - Clean the reinforcement steel of any corrosion.
   - Coat the reinforcement steel and the concrete surface with an anti-corrosion coating.

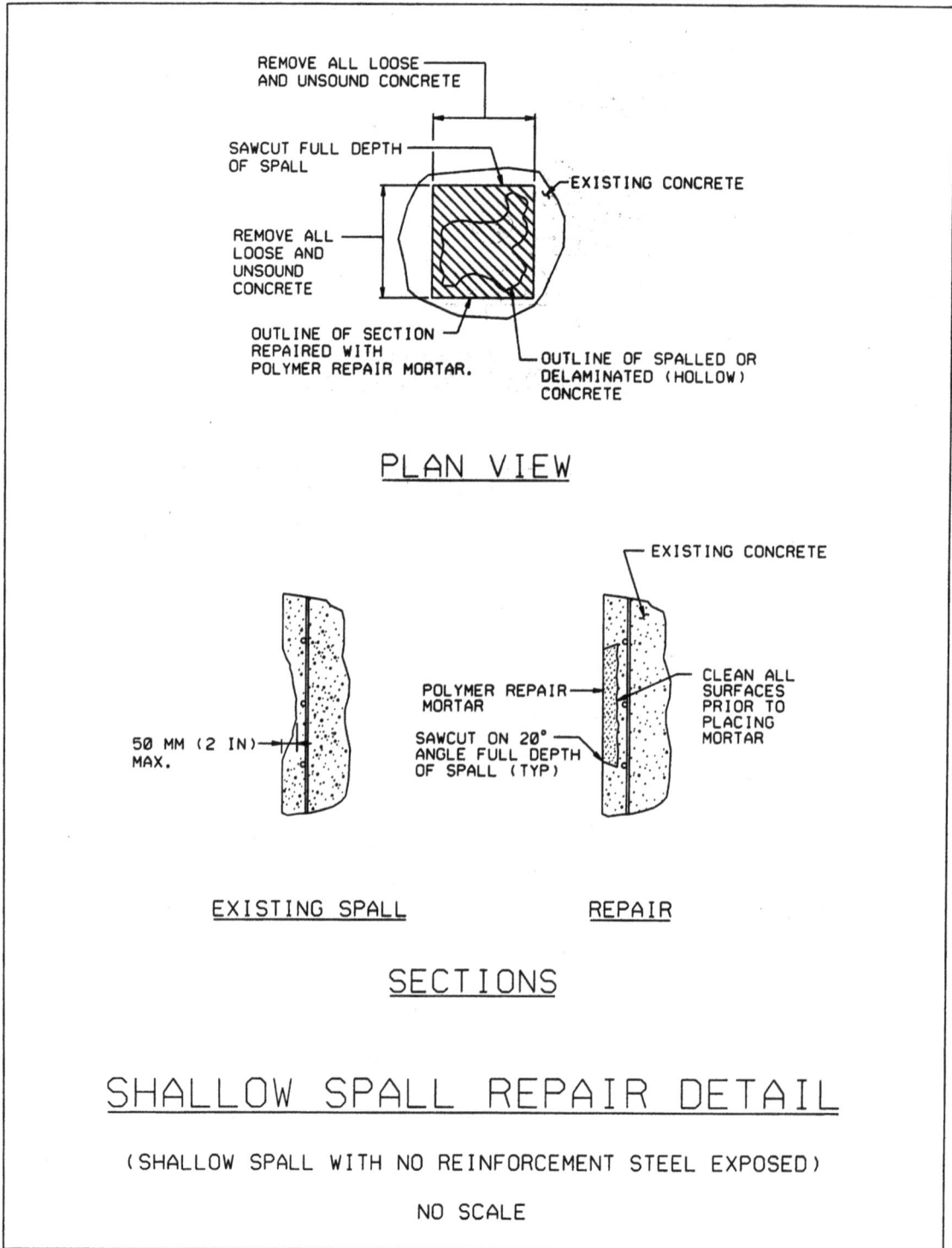

REMOVE ALL LOOSE
AND UNSOUND CONCRETE

SAWCUT FULL DEPTH
OF SPALL

EXISTING CONCRETE

REMOVE ALL
LOOSE AND
UNSOUND
CONCRETE

OUTLINE OF SECTION
REPAIRED WITH
POLYMER REPAIR MORTAR.

OUTLINE OF SPALLED OR
DELAMINATED (HOLLOW)
CONCRETE

PLAN VIEW

50 MM (2 IN)
MAX.

EXISTING CONCRETE

POLYMER REPAIR
MORTAR

SAWCUT ON 20°
ANGLE FULL DEPTH
OF SPALL (TYP)

CLEAN ALL
SURFACES
PRIOR TO
PLACING
MORTAR

EXISTING SPALL   REPAIR

SECTIONS

SHALLOW SPALL REPAIR DETAIL

(SHALLOW SPALL WITH NO REINFORCEMENT STEEL EXPOSED)

NO SCALE

Figure 4.13

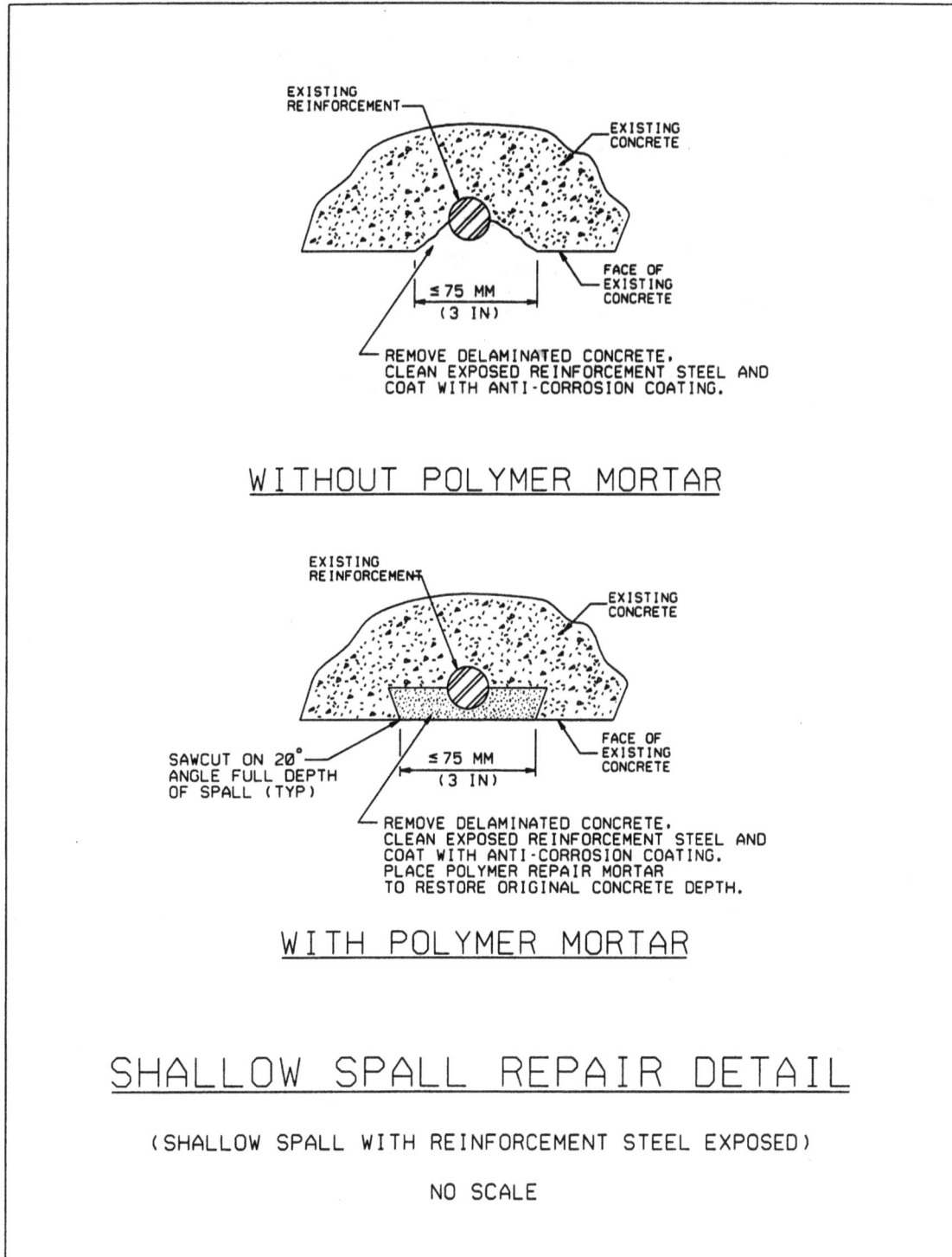

EXISTING
REINFORCEMENT

EXISTING
CONCRETE

FACE OF
EXISTING
CONCRETE

≤ 75 MM
(3 IN)

REMOVE DELAMINATED CONCRETE,
CLEAN EXPOSED REINFORCEMENT STEEL AND
COAT WITH ANTI-CORROSION COATING.

WITHOUT POLYMER MORTAR

EXISTING
REINFORCEMENT

EXISTING
CONCRETE

SAWCUT ON 20°
ANGLE FULL DEPTH
OF SPALL (TYP)

≤ 75 MM
(3 IN)

FACE OF
EXISTING
CONCRETE

REMOVE DELAMINATED CONCRETE,
CLEAN EXPOSED REINFORCEMENT STEEL AND
COAT WITH ANTI-CORROSION COATING.
PLACE POLYMER REPAIR MORTAR
TO RESTORE ORIGINAL CONCRETE DEPTH.

WITH POLYMER MORTAR

SHALLOW SPALL REPAIR DETAIL

(SHALLOW SPALL WITH REINFORCEMENT STEEL EXPOSED)

NO SCALE

Figure 4.14

- If replacing the spalled concrete is recommended, then, prior to application of anti-corrosion coating, perform sawcut as described in Part a) and place polymer repair mortar as final step. Make sure that anti-corrosion coating and polymer repair mortar are chemically compatible.

c)   Deep Spall With Reinforcement Steel Exposed (See Figures 4.15 and 4.16)

Generally, any exposed reinforcement steel in a deep spall will be corroded. The extent of this corrosion should be determined and the concrete should be removed around the effected reinforcement steel to a width of a least one half the existing reinforcement steel spacing and to a depth of at least 25 mm (1 in) behind the back of the reinforcement steel. It is recommended that the sawcut around the perimeter of the spalled area be at least 25 mm (1 in) deep to accommodate a repair material with aggregate. If the material being used does not include aggregate, that depth can be reduced to 6 mm (¼ in), given that a proper bonding agent is used. As for bonding agents, experience has shown that separate, manual application is often not performed correctly and insufficient coverage is obtained. Therefore, a bonding agent admixture can be substituted for a certain percentage of the water in the mix. Specific repair recommendations are as follows:

- Remove all loose or delaminated concrete from the spalled surface and face of the reinforcement steel.
- Clean the concrete and steel surfaces of deleterious materials.
- Sawcut around the spalled area.
- Provide new reinforcement steel where necessary and overlap with existing steel according to current American Concrete Institute (ACI) standards.
- Coat the reinforcement steel with an anti-corrosion coating.
- Place polymer repair mortar in the spalled area unless the area is very large such that the use of shotcrete or plasticized concrete pumped with a form is more cost-effective. Where shotcrete is used, additional welded wire fabric is recommended to help support the shotcrete.

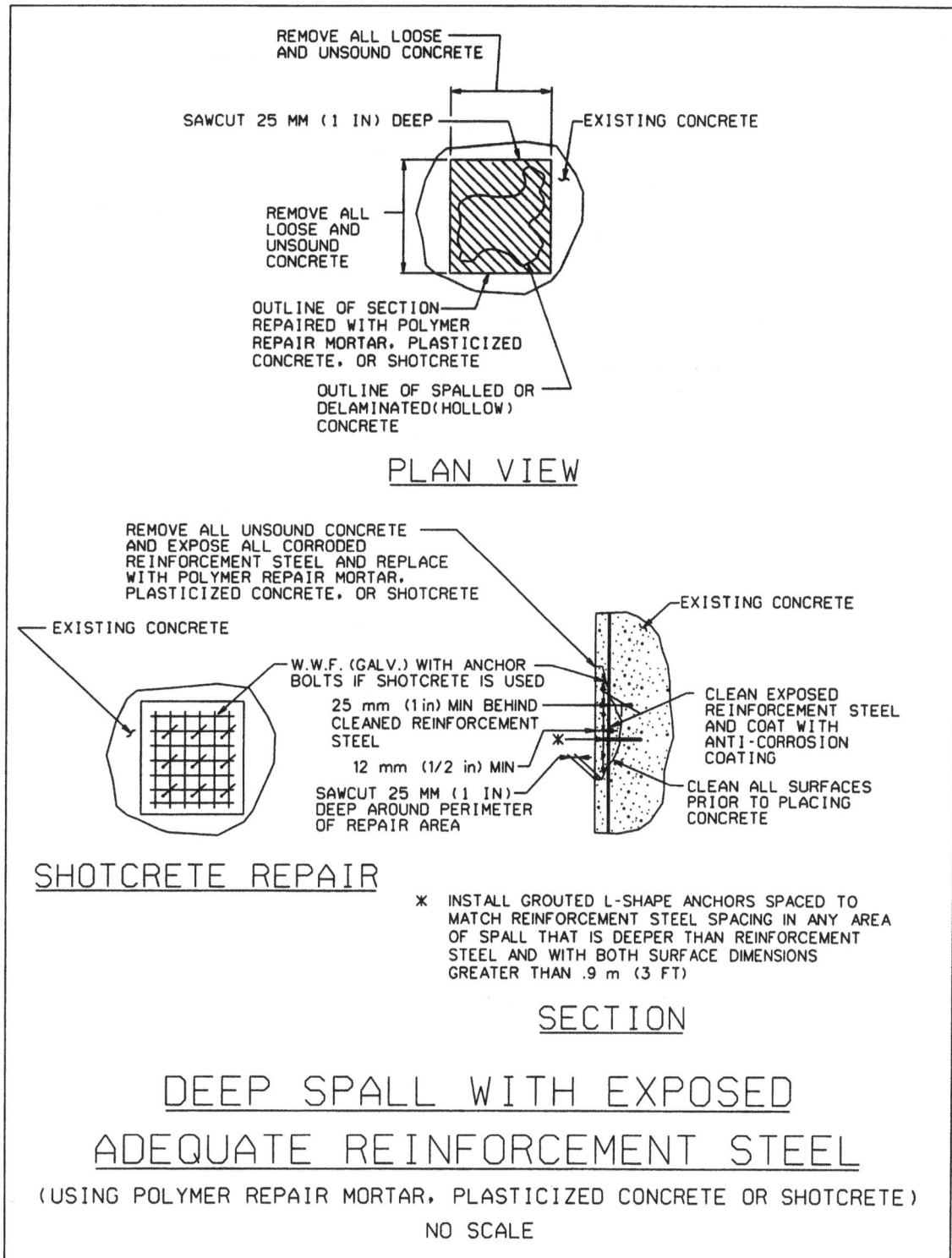

REMOVE ALL LOOSE
AND UNSOUND CONCRETE

SAWCUT 25 MM (1 IN) DEEP

EXISTING CONCRETE

REMOVE ALL
LOOSE AND
UNSOUND
CONCRETE

OUTLINE OF SECTION
REPAIRED WITH POLYMER
REPAIR MORTAR, PLASTICIZED
CONCRETE, OR SHOTCRETE

OUTLINE OF SPALLED OR
DELAMINATED(HOLLOW)
CONCRETE

PLAN VIEW

REMOVE ALL UNSOUND CONCRETE
AND EXPOSE ALL CORRODED
REINFORCEMENT STEEL AND REPLACE
WITH POLYMER REPAIR MORTAR,
PLASTICIZED CONCRETE, OR SHOTCRETE

EXISTING CONCRETE

EXISTING CONCRETE

W.W.F. (GALV.) WITH ANCHOR
BOLTS IF SHOTCRETE IS USED

25 mm (1 in) MIN BEHIND
CLEANED REINFORCEMENT
STEEL

12 mm (1/2 in) MIN

CLEAN EXPOSED
REINFORCEMENT STEEL
AND COAT WITH
ANTI-CORROSION
COATING

SAWCUT 25 MM (1 IN)
DEEP AROUND PERIMETER
OF REPAIR AREA

CLEAN ALL SURFACES
PRIOR TO PLACING
CONCRETE

SHOTCRETE REPAIR

✱ INSTALL GROUTED L-SHAPE ANCHORS SPACED TO
MATCH REINFORCEMENT STEEL SPACING IN ANY AREA
OF SPALL THAT IS DEEPER THAN REINFORCEMENT
STEEL AND WITH BOTH SURFACE DIMENSIONS
GREATER THAN .9 m (3 FT)

SECTION

DEEP SPALL WITH EXPOSED
ADEQUATE REINFORCEMENT STEEL
(USING POLYMER REPAIR MORTAR, PLASTICIZED CONCRETE OR SHOTCRETE)
NO SCALE

Figure 4.15

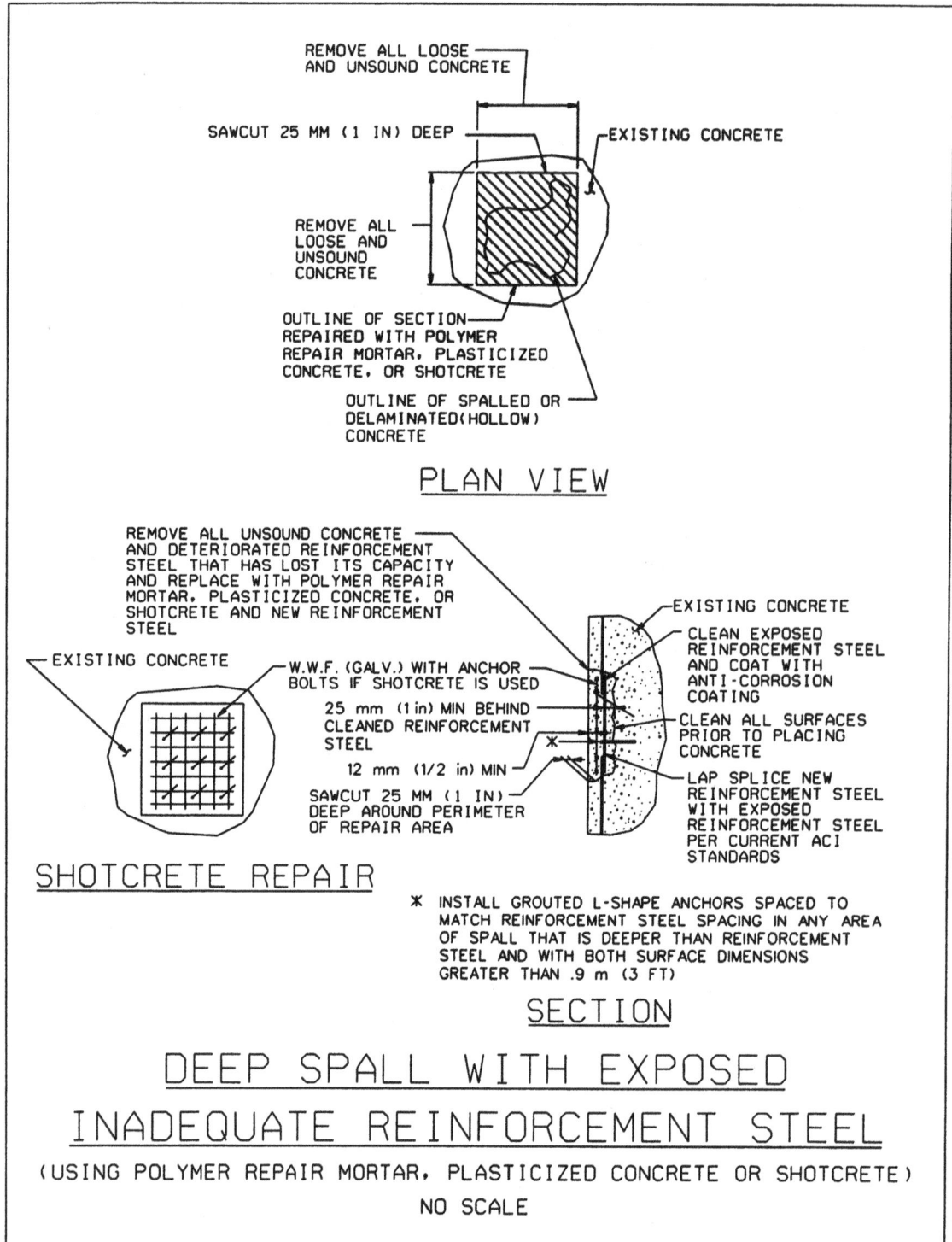

REMOVE ALL LOOSE
AND UNSOUND CONCRETE

SAWCUT 25 MM (1 IN) DEEP

EXISTING CONCRETE

REMOVE ALL
LOOSE AND
UNSOUND
CONCRETE

OUTLINE OF SECTION
REPAIRED WITH POLYMER
REPAIR MORTAR, PLASTICIZED
CONCRETE, OR SHOTCRETE

OUTLINE OF SPALLED OR
DELAMINATED(HOLLOW)
CONCRETE

PLAN VIEW

REMOVE ALL UNSOUND CONCRETE
AND DETERIORATED REINFORCEMENT
STEEL THAT HAS LOST ITS CAPACITY
AND REPLACE WITH POLYMER REPAIR
MORTAR, PLASTICIZED CONCRETE, OR
SHOTCRETE AND NEW REINFORCEMENT
STEEL

EXISTING CONCRETE

W.W.F. (GALV.) WITH ANCHOR
BOLTS IF SHOTCRETE IS USED

25 mm (1 in) MIN BEHIND
CLEANED REINFORCEMENT
STEEL

12 mm (1/2 in) MIN

SAWCUT 25 MM (1 IN)
DEEP AROUND PERIMETER
OF REPAIR AREA

EXISTING CONCRETE

CLEAN EXPOSED
REINFORCEMENT STEEL
AND COAT WITH
ANTI-CORROSION
COATING

CLEAN ALL SURFACES
PRIOR TO PLACING
CONCRETE

LAP SPLICE NEW
REINFORCEMENT STEEL
WITH EXPOSED
REINFORCEMENT STEEL
PER CURRENT ACI
STANDARDS

SHOTCRETE REPAIR

✽ INSTALL GROUTED L-SHAPE ANCHORS SPACED TO
MATCH REINFORCEMENT STEEL SPACING IN ANY AREA
OF SPALL THAT IS DEEPER THAN REINFORCEMENT
STEEL AND WITH BOTH SURFACE DIMENSIONS
GREATER THAN .9 m (3 FT)

SECTION

DEEP SPALL WITH EXPOSED
INADEQUATE REINFORCEMENT STEEL

(USING POLYMER REPAIR MORTAR, PLASTICIZED CONCRETE OR SHOTCRETE)

NO SCALE

Figure 4.16

C.   **LINER REPAIRS**

A general note that applies to all the liner repairs suggested in the following sections, is that a registered professional engineer should evaluate and approve suggested repairs and methods used. Particular attention should be directed to determining if structural components need to be temporarily shored so that the component to be repaired is unloaded.

1.   **Cast-in-Place (CIP) Concrete**

CIP concrete liners are common in both highway and transit tunnels because of strength, cost, adaptability to site conditions, durability and resistance to corrosion (if designed and constructed properly), and ability to obtain a smooth surface for the final tunnel finish (tile, metal panels, etc.) application. Although there are many benefits for using CIP concrete liners, they may also have extensive repair needs to remedy cracking, spalling, and reinforcement steel deterioration. These effects could be due to water infiltration, inadequate design/construction, age, or unforeseen changes in ground conditions surrounding the tunnel.

The methods for repair of CIP concrete liners are the same as those given for general concrete in Chapter 4, Sections A and B, but will briefly be reiterated below for reference.

- • If water infiltration is occurring, then methods of water redirection, crack injection, soil grouting, or membrane application should be performed prior to actual concrete repair.
- • If dry cracks need structural repair, epoxy resins can be injected, but a determination must be made if there are active movements at the crack. If actively moving cracks are epoxy grouted, then subsequent cracks adjacent to original crack may occur depending on the elastic capacity of the epoxy material. Other materials with cellular structures can be used for active cracks.
- • Spall repair is dependent on the size and depth of the spall and can be repaired with a polymer mortar for smaller spalls or with a plasticized concrete or shotcrete for larger spalls. Care must be given to cleaning or replacing exposed steel that has experienced corrosion and section loss.

Oftentimes in highway tunnels, the CIP concrete liner is covered with a reflective material such as tile or metal panels; therefore, the repair technique must take into account the attachment requirements of the final tunnel finish.

Another retrofit method that is being used more often for strengthening concrete tunnel linings is carbon fiber, polyaramide glass fiber sheet products. This method is performed by first completing crack and spall repairs, and then the concrete surface is prepared as per the manufacturer's instructions. An epoxy coating is applied to the concrete surface and the fiber sheets are installed in two layers, the sheets with fibers in the transverse (circumferential) direction are installed first followed by sheets with fibers in the longitudinal direction. The sheets are impregnated with epoxy.

This type of retrofit can be used to increase the load capacity of the tunnel arch when there is increase in the weight of overburden, this will also help to prevent further cracking and exfoliation of the concrete lining. **Caution:** This method should not be used in areas where fires or excessive heat may occur, due to the possible flammability and toxicity of the materials used. In the event of a fire, these materials will fail and therefore the concrete lining will lose any structural improvements provided by the carbon fiber sheets. Also, this method is not recommended in areas of the tunnel that might experience water infiltration.

2. Precast Concrete

Precast concrete liners are often used as a primary liner that is placed by the TBM or manually within the shield of a driven tunnel. They are used because of their easy adaptability to site conditions, and speed of erection. In highway tunnels, the precast concrete liners are often covered with an interior cast-in-place concrete liner for supporting the tunnel finish as described previously. Conversely, transit tunnels, which do not have the same visibility constraints, sometimes use a single precast concrete liner with no interior finish; therefore repairs are made to the precast directly. Generally precast segmental liners are bolted together to compress gaskets in the joints to prevent water infiltration and to provide overall structural stability to the liner.

Repair of precast segmental concrete liners is often related to degradation of the joints, especially in tunnels subject to water infiltration. The joint material can fail and corrosion of the bolts can lead to spalling of the concrete, which can expose the reinforcement steel, subsequently subjecting it to corrosion effects as well. Obviously, the ideal is to repair the joint before the corrosion becomes too extensive; therefore a routine inspection is crucial. As mentioned previously in the water infiltration section, the method of repairing a joint consists of repacking the joint with new gasket material and replacing any bolts that have lost their structural capacity. Also, the joints can be injected with grout to help seal them off to water.

Other defects such as cracks and spalls that can occur within a precast panel can be repaired using the same methods given in either Chapter 4, Section A or B depending on whether water infiltration is present.

3. Steel

Within a tunnel, structural steel is used for two main purposes: as segmental steel liners and as structural columns or beams. Structural columns and beams are mostly found in transit tunnels although steel beams are also used in structural slabs for support of roadway or overlying buildings and tunnels. As with structural steel in other uses such as bridges and buildings, the primary method of failure is by corrosion caused by moisture, which in transit tunnels can be enhanced by the presence of stray current from the rail electrification system. It is also possible for steel to develop cracks due to improper design/erection, fatigue, or from defects in the material.

To repair these defects, it is necessary to determine the cause and actual extent of the damage. Typically this will be done during the inspection process and will be recorded for reference in determining the type of repair. One general note is that for older structures the weldability of the steel must be determined due to the wider range of chemical composition allowed in their fabrication. Table 4.1 illustrates the changes in steel weldability over time.

Table 4.1 – Weldability of Steel

Dates	Weldability
Prior to 1923	Steel should be tested
1923-1936	Generally weldable
After 1936	Weldable

If the existing steel utilizes welded connections, then it is safe to assume that the steel is weldable. But, if there is any doubt, then the steel should be tested according to American Welding Society (AWS) standards.

Below are examples of repair procedures that can be used for steel defects:

- If beams or columns made from W-shapes, T-shapes, or channels have significant section loss (greater than 20 percent), then consider welding or bolting plates to flanges or webs to increase the capacity in the area of the section loss.
- For steel segmental liners that have section loss or considerable corrosion of the panels, then plates can be welded on the interior surface to replace the area of section loss.
- If liner joints and bolts are corroded, then new joint material must be installed along with new bolts. If stray current is suspected, then install an insulating sleeve over the bolt to prevent current from passing between dissimilar metals.
- Painting the steel is the best method for preventing corrosion. Research should be conducted to determine the best paint type for the given situation. Traditionally, epoxy paints have performed well for steel. Prior to painting, existing steel should be blast cleaned of all present corrosion – down to white metal.
- If clearance is adequate, headed studs may be welded to the liner and then a layer of reinforced concrete or shotcrete can be constructed inside the steel liner. If welding is not practical, threaded rods that anchor the re-bars may replace steel bolts that connect the liner segments.

4.   **Cast Iron**

Cast iron is similar to steel in the extent of its use for tunnel construction, such as the primary tunnel liner segments and columns (usually tubular) in open areas of transit tunnels. Cast iron differs from steel in that it is not as susceptible to corrosion. Generally cast iron is far less ductile than steel and therefore brittle failure and cracking can be more common.

Repair of cast iron defects is much more difficult than for steel and therefore a detailed, site-specific investigation is required to determine the proper method for repair. However, there are some general comments that can be made about repair methods that can be used.

a)   Bolting

It is possible to bolt new cast iron members over existing cracks, or areas of corrosion. When doing so, a watertight connection must be accomplished. If the repair is at a joint between liner segments, then the joint itself could be made watertight by inserting gasket material or by injecting the joint with a chemical grout. If the repair is the addition of a plate over a crack or area of section loss in the panel, then a waterproofing material will need to be applied between the new piece and the existing lining.

b)   Welding

In general cast iron that is used in tunnels should **not** be considered weldable. Depending on the type of cast iron (gray, nodular, white, malleable, etc.) and the accessibility of the item to be repaired, some welding techniques can be attempted. Significant expertise is required and preheating is necessary, which is difficult since the cast iron components in a tunnel are not usually removable. Therefore, other methods of repair will usually be recommended.

c)   Concrete Liner

Similarly to steel liners, if clearance is adequate, a layer of reinforced concrete or shotcrete can be added inside the cast iron liner. However as mentioned above welding is not usually an option, so replacing bolts or rivets that connect the liner segments with threaded rods to anchor the re-bar is suggested.

d)   Metal Stitching

Technology does exist to stitch the cast iron in a manner illustrated in Figures 4.17 – 4.19. It is recommended that if the cast iron cannot be repaired using other methods, that this method be investigated. Currently, this method is being used with much success on high-pressure castings such as water pumps, valves, compressors and pipes, which demonstrate that the method provides a high strength, watertight repair. This process can restore the original strength to the casting without the problems associated with on-site welding such as stress, distortion, hardening and additional cracking because heat is not used in the repair process.

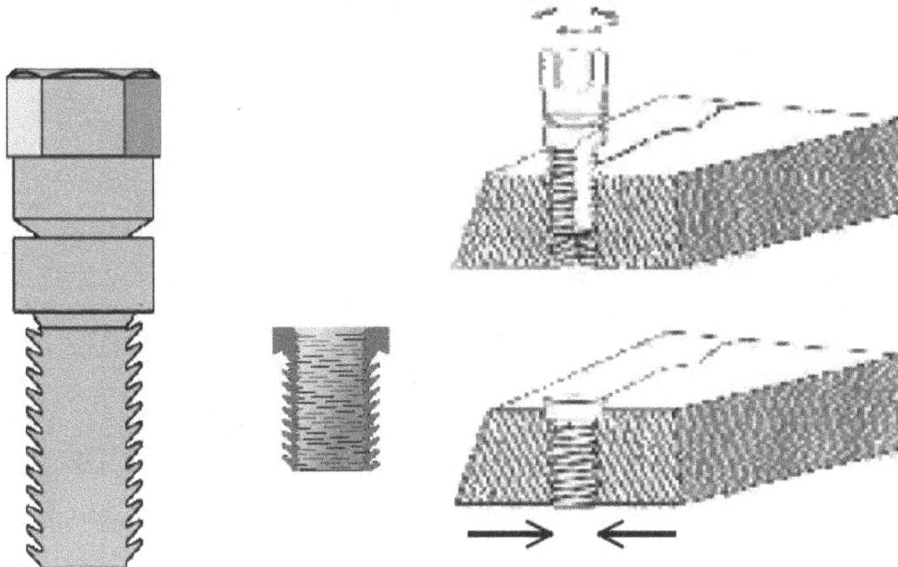

Figure 4.17 – Metal Stitching Detail – (Figure courtesy of Lock-N-Stitch Inc.)

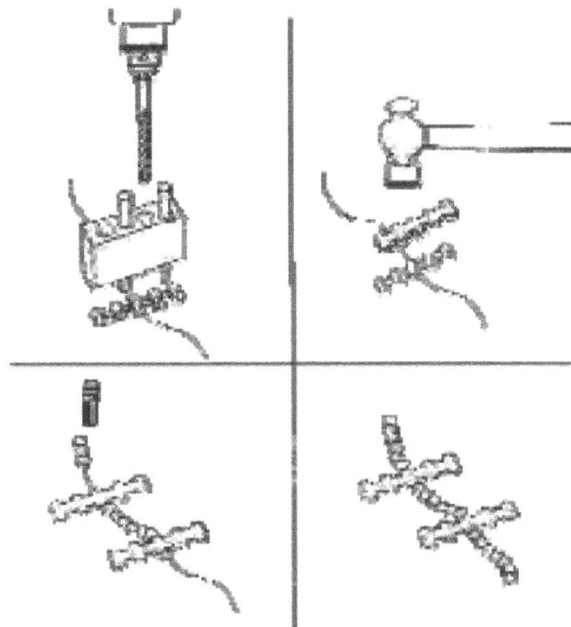

Figure 4.18 – Metal Stitching Procedure – (Figure courtesy of Lock-N-Stitch Inc.)

Figure 4.19 – Metal Stitching Completed – (Photo courtesy of Lock-N-Stitch Inc.)

5.   Shotcrete

Shotcrete is a material that is gaining increasing usage for tunnel construction as materials and methods of application continually improve. Another terminology that is sometimes used is "gunite," which refers to fine-aggregate shotcrete. There are various uses for shotcrete in tunnel construction and each use may require a different mix design and application method. Primarily it is used as a primary support liner for the excavation prior to the construction of the final liner. This procedure can be supplemented with rock bolts, lattice girders, or wire mesh for additional strength. More recently with the addition of steel or synthetic fibers and fine-aggregates, shotcrete has been able to be used as a final liner, which can achieve significant strength in thin, smooth layers. Shotcrete can be used to cover and protect a waterproofing liner or as a repair liner for tunnel rehabilitation.

Generally, cured shotcrete will behave similarly to standard cast-in-place concrete and will be susceptible to cracking, spalling and delamination, even though the mix designs were intended to reduce those effects. If repairs need to be made to shotcrete liners, they can be performed in the same manner as the methods given in Chapter 4, Section A and B, depending on whether water is present at the defect.

6.   Masonry

The term "masonry" refers to materials such as stone or brick that are connected together in the field with mortar, which in the case of brick tunnel liners could be five or more courses thick. In older tunnels–generally those built in the 19th century–masonry was the construction material that was most readily available and economically possible for construction of the liners. Oftentimes, even after concrete and steel began to be used as construction materials for cut-and-cover tunnels, masonry was still used as a protective liner for the mastic waterproofing that was used on the outside of the finished lining. Masonry liners that are still in existence today range from very good condition to very poor condition, depending on the severity of any ground water presence. If they were constructed within geologic conditions that kept them relatively free from the presence of ground water, the masonry itself could last without much attention for a very long time. This is proven by the fact that most of the world's historic tunnel structures were constructed with masonry and still exist today.

Another reason that masonry tunnel liners remain in good condition is that the original method for waterproofing against, or draining of, the ground water was and remains very effective. The original waterproofing system typically consisted of a timber primary support lining and void space between the timber and masonry that was filled with tunnel debris, which formed a drainage channel for ground water. Over time, the timber lining rots and the water erodes the material that filled the void, causing the masonry itself to be exposed to the water.

When masonry is exposed to water, it and the mortar can swell and become brittle depending on the firing temperature of the brick and the chemical make up of the mortar. This, in conjunction with possible ground collapse in the space behind the brick lining, can induce stresses into the lining that cannot be resisted and therefore, structural cracking occurs, which further exacerbates the water infiltration problem.

If it is determined that repairs are needed, then the actual cause of the deficiency needs to be determined in order to select the proper repair method. If the problem is caused by extensive water infiltration, then methods given in Chapter 4, Section A should be considered, otherwise the following are suggested:

- Inject cementitious grout into known large voids behind liner to stabilize ground material. If waterproofing is needed, use methods described in Chapter 4, Section A, Part 3(b)(4).
- Replace cracked or brittle masonry units in localized areas.
- Repoint mortar by removing existing mortar to depth of twice the joint thickness or 18 mm (¾ in) minimum and replacing with new mortar of equal strength and color but increased water impermeability.
- Provide horizontal reinforcement steel embedded in the joint across the crack prior to repointing, for added strength.
- Inject cracks with chemical or particle grouts (take care to use grouts that are suited for the moisture content present in the crack).
- Apply a shotcrete lining if vertical and horizontal clearances can be reduced. However, underlying causes of cracks and water infiltration must be addressed first.

One repair that is **not** recommended is to apply an impermeable coating–such as a paint or epoxy–to the interior surface of the masonry. This practice is discouraged because any moisture or water that enters the masonry will be trapped and cause swelling; inevitably the face of the masonry will delaminate and fall off.

7.   **Exposed Rock**

Many older tunnels that were constructed through dry, sound rock conditions, were left unlined except for zones near the portals or where the rock was incompetent to carry the loads. These tunnels may function without need for repair long into the future, but it is more likely that ground movements will either cause pieces of rock to fracture and fall to the invert, thus endangering the tunnel occupants, or they will open up cracks in which water will eventually infiltrate into the tunnel space. For the latter situation, some type of waterproofing liner, membrane, or pipe network will most likely need to be installed at the location of the leak to divert the water towards the tunnel drainage system.

If water infiltration is not a concern, then there are methods that can be used to structurally support the exposed rock, so that it does not pose a threat to the tunnel occupants. Listed below and shown in Figure 4.17 are some examples of those methods:

- Metal plates attached with short anchor bolts can be used to support surface defects. The plates can vary in width and length as needed to cover fractured rock.
- Rock bolts can be used to secure thicker sections of fractured rock to a competent layer behind. Examples of rock bolts are standard rock bolts, cable bolts, or friction bolts (dowels). They can sometimes be prestressed, but normally the stress is induced during future ground movements. Also, they are normally grouted into the drilled hole using chemical or particle grouts, but can be anchored mechanically for short-term applications.
- To protect from small spalls or pieces of fractured rock, wire mesh (chain link) can be attached to the surface using rock bolts.
- To completely protect against falling debris and to increase the structural capacity of the liner, shotcrete or a thin cast-in-place liner can be installed. However, this process does reduce the interior clearances.

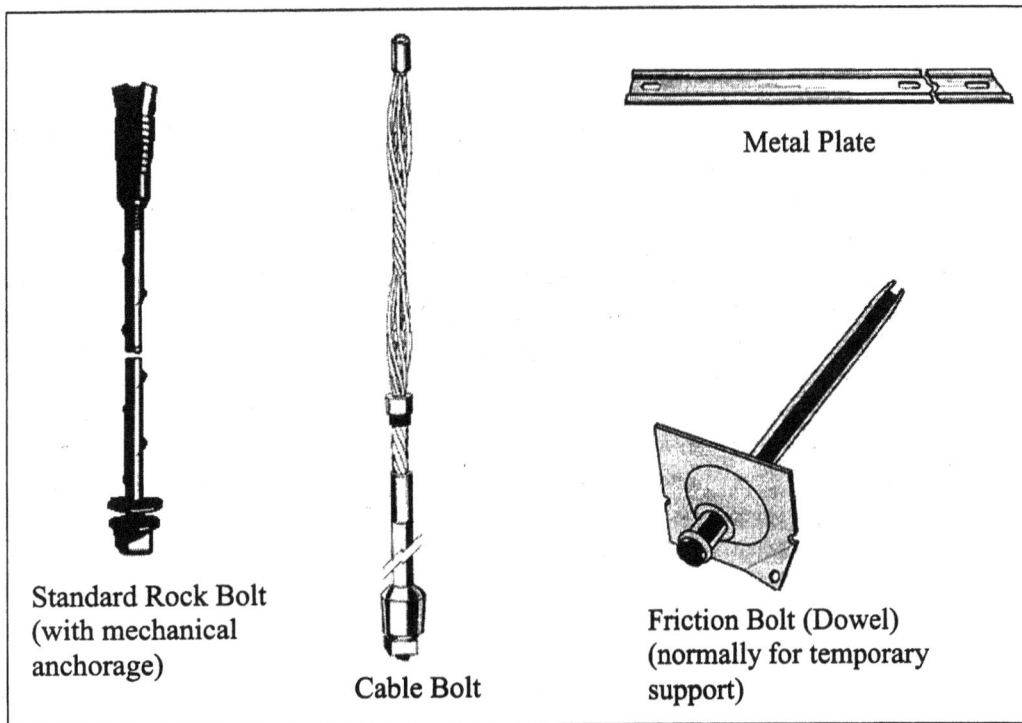

Figure 4.20 – **Rock Bolt Types**

APPENDIX A:
LIFE-CYCLE COST METHODOLOGY

To properly plan for future repairs or scheduled maintenance in a tunnel, it is beneficial to perform a life cycle cost analysis of the different options involved for each anticipated major repair to ensure the greatest cost efficiency over the life of the tunnel. This process involves evaluating the alternatives over a given duration or economic life to determine specific costs involved for each option and then equating them through a series of mathematical formulas that enable the costs of each option to be compared at a common point in time. The life cycle costs of a given alternative include all associated costs over the expected life of the option. In general these costs may include:

- Initial costs – Engineering or design costs, price of equipment, construction costs, etc.
- Operating/energy costs – Annualized amount to operate (e.g. cost of electricity to run mechanical equipment).
- Maintenance costs – Annualized costs to maintain equipment or repair minor defects.
- Rehabilitation costs – Future expense for known procedure at specified time (e.g. certain type of light bulbs may need to be replaced every five years).
- User costs – Costs associated with impact on the functioning of tunnel (e.g. tunnel may need shut down for repair; therefore, impact to traffic can be shown by applying an annualized cost to each hour tunnel is closed).
- Salvage value – Sale value of equipment at end of expected life (e.g. mechanical fan or railroad tie may be of some value to others even after it has served its purpose in the tunnel).

There are two main methods for performing a life cycle cost analysis, namely, the present worth and the annualized methods.

1. Present Worth Method

As the name implies, this method attempts to bring all of the present and future costs of a given option to present day values. This process should be completed for each major repair/rehabilitation and subsequently the options could be compared. Determining the present worth of a future expense is done by taking into account inflation of the dollar and therefore discounting the amount by a predetermined rate over the period between the future expense and the present time. The present worth of the future expense is also the amount that could be invested today with reinvested interest over the duration to equal the amount of the future expense. An example of a future expense would be the rehabilitation costs mentioned above. The general form of the equation for determining the present worth of a future expense is:

$$P = F \left\{ \frac{1}{(1+i)^n} \right\}$$

Where P =   Present worth
F =   Future one-time expense
n =   Number of years
i =   Discount rate

Future expenses can also be uniform, in that the same expense occurs at the end of each year. An example of this would be the annualized maintenance costs described previously. The general form of the equation for determining the present worth of an end-of-year expense is:

$$P = A \left\{ \frac{[(1+i)^n - 1]}{[i(1+i)^n]} \right\}$$

Where P =   Present worth
A =   End-of-year payments
n =   Number of years
i =   Discount rate

2. Annualized Method

The annualized method is used to transform present and future costs into a uniform annual expense. This annual expense can be compared to the annual expenses of the other repair/rehabilitation alternatives to determine which one is most cost effective. Converting all future expenses into a present value as before and then using the equation below to convert that value into an annual expense will provide a uniform annual cost.

$$A = P \left\{ \frac{[i(1+i)^n]}{[(1+i)^n - 1]} \right\}$$

Where A =   End-of-year payments
P =   Present worth
n =   Number of years
i =   Discount rate

The above two methods can also be performed without using the actual equations given. Standard economic tables have been developed that give factors that are based on the discount rate and the economic life under consideration. These factors are also unique to the desired result. The procedure for using standard economic tables is as follows:

- Determine the discount rate (i) and economic life (n) to be used for the analysis. It is important to choose an economic life that is equal for the given alternatives if the present worth method is to be used. Otherwise, the annualized method must be used.

- Develop a cash flow diagram for each option, which shows all relevant costs described above on a timeline of years in the economic life.

- Take individual costs, whether uniform or one-time, and insert them in the proper formula given below along with the factor from the appropriate economic table.

  (P/F,i%,n) – or present worth *(P)* given future expense *(F)* at discount rate *(i)* for number of years *(n)*.

  (P/A,i%,n) – or present worth *(P)* given end-of-year payments *(A)* at discount rate *(i)* for number of years *(n)*.

  (A/P,i%,n) – or end-of-year payments *(A)* given present worth *(P)* at discount rate *(i)* for number of years *(n)*.

Caution must be used in determining the appropriate discount rate. Because of the power of compounded interest, a difference in discount rate can actually change the final outcome of the analysis if the repair/rehabilitation options being considered have different arrangements of uniform and onetime costs. According to Peter Kleskovic who wrote *A Discussion of Discount Rates for Economic Analysis of Pavements*, a draft report for FHWA:

"The discount rate can affect the outcome of a life cycle cost analysis in that certain alternatives may be favored by higher or lower discount rates. High discount rates favor alternatives that stretch out costs over a period of time, since the future costs are discounted in relation to the initial cost. A low discount rate favors high initial cost alternatives since future costs are added in at almost face value. In the case of a discount rate equal to 0, all costs are treated equally regardless of when they occur. Where alternative strategies have similar maintenance, rehabilitation, and operating costs, the discount rate will have a minor effect on the analysis and initial costs will have a larger effect."

The above procedures will allow the most economical repair/rehabilitation alternative to be identified, but as can be expected, the least costly is not always the best. Therefore further comparison can sometimes be utilized to take into account the "human factors" of the alternatives. In his book entitled *Value Engineering: Practical Applications ... for Design, Construction, Maintenance & Operations*, Alphonse Dell'Isola, P.E., has developed a procedure for weighted evaluation of human factors such as comfort, appearance, performance, and safety along with the economic costs. It is suggested that his procedure or something similar be used if the effects of the human factors are of concern during the economic life of the alternatives.

3.   Example

The following example uses completely arbitrary costs to properly show the benefits of a life cycle cost analysis.

Consider a transit tunnel in which the track support system is in need of replacement. Currently the system is ballasted track and can either be replaced with direct fixation slab track or a new, ballasted track system. Costs given to the different options are shown below for every 150 m (500 ft) of track.

a)   Direct Fixation Slab Track:

Initial Construction Costs $500,000
Joint/Crack Sealing (years 10, 20, 30 and 40) $20,000
Annual Maintenance $1,000
Salvage.. ($100,000)
Estimated Life.. 50 years

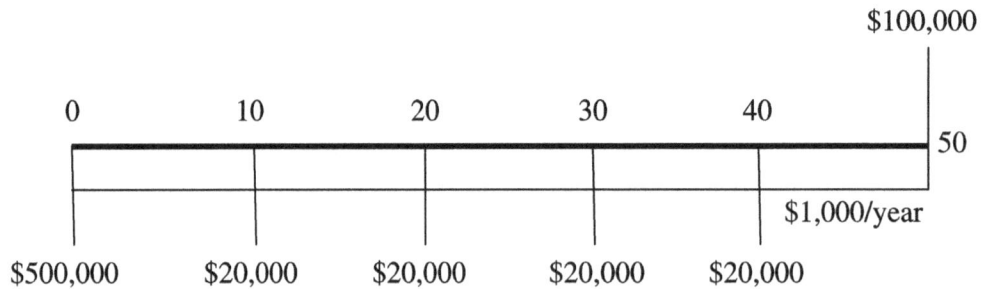

b)   Ballasted Track:

Initial Construction Costs $250,000
Replacement Ties (years 12 and 24) $200,000
Annual Maintenance $20,000
Salvage.. ($50,000)
Estimated Life.. 35 years

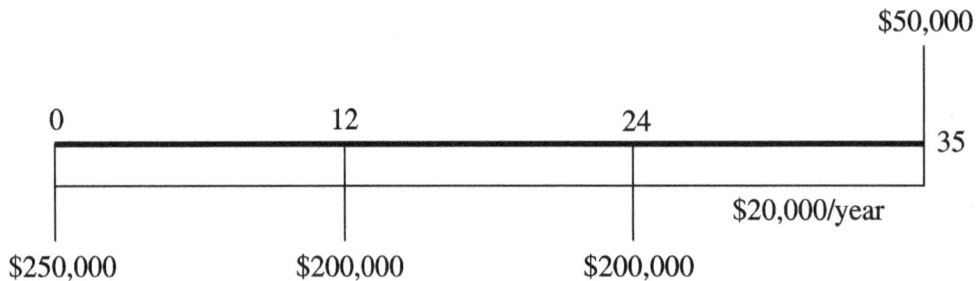

c)   Factors from Standard Economic Table (assume 7 percent discount rate)

n	(P/F)	(P/A)	(A/P)
10	0.5083	7.0236	0.1424
12	0.4440	7.9427	0.1259
20	0.2584	10.5940	0.0944
24	0.1971	11.4693	0.0872
30	0.1314	12.4090	0.0806
35	0.0937	12.9477	0.0772
40	0.0668	13.3317	0.0750
50	0.0339	13.8007	0.0725

d)   Alternate 1 – Direct Fixation Slab Track

(1)   Present Worth Method

$$P = \$500,000 + \$1,000(P/A,7\%,50) + \$20,000(P/F,7\%,10) + \$20,000(P/F,7\%,20) + \$20,000(P/F,7\%,30) + \$20,000(P/F,7\%,40) - \$100,000(P/F,7\%,50)$$

$$P = \$500,000 + \$1,000(13.8007) + \$20,000(0.5083) + \$20,000(0.2584) + \$20,000(0.1314) + \$20,000(0.0668) - \$100,000(0.0339)$$

$$P = \$529,709$$

(2)   Annualized Method

$$A = \$500,000(A/P,7\%,50) + \$1,000 + \$20,000(P/F,7\%,10)(A/P,7\%,50) + \$20,000(P/F,7\%,20)(A/P,7\%,50) + \$20,000(P/F,7\%,30)(A/P,7\%,50) + \$20,000(P/F,7\%,40)(A/P,7\%,50) - \$100,000(P/F,7\%,50)(A/P,7\%,50)$$

$$A = \$500,000(0.0725) + \$1,000 + \$20,000(0.5083)(0.0725) + \$20,000(0.2584)(0.0725) + \$20,000(0.1314)(0.0725) + \$20,000(0.0668)(0.0725) - \$100,000(0.0339)(0.0725)$$

$$A = \$38,403/\text{year}$$

e)   Alternate 2 – Ballasted Track

         (1)   Present Worth Method

$$P = \$250,000 + \$20,000(P/A,7\%,35) +$$
$$\$200,000(P/F,7\%,12) +$$
$$\$200,000(P/F,7\%,24) -$$
$$\$50,000(P/F,7\%,35)$$

$$P = \$250,000 +$$
$$\$20,000(12.9477) +$$
$$\$200,000(0.444) +$$
$$\$200,000(0.1971) -$$
$$\$50,000(0.0937)$$

$$P = \$631,689$$

         (2)   Annualized Method

$$A = \$250,000(A/P,7\%,35) +$$
$$\$20,000 +$$
$$\$200,000(P/F,7\%,12)(A/P,7\%,35) +$$
$$\$200,000(P/F,7\%,24)(A/P,7\%,35) -$$
$$\$50,000(P/F,7\%,35)(A/P,7\%,35)$$

$$A = \$250,000(0.0772) +$$
$$\$20,000 +$$
$$\$200,000(0.444)(0.0772) +$$
$$\$200,000(0.1971)(0.0772) -$$
$$\$50,000(0.0937)(0.0772)$$

$$A = \$48,837/\text{year}$$

Since this example used two different time periods the present worth results are not useful in comparing costs, but the annualized method is since the outcome is a cost per year. From this example it can be seen that even though Alternative 2 had a lower initial cost, it turned out to be more expensive over the long term due to greater costs to repair and maintain that alternative. This process can be used in evaluating many different aspects of tunnel maintenance and repairs from structural aspects like the example above to fan model selection for the mechanical ventilation system to which light bulb manufacturer is better over the long term. There are multiple reference materials that could be of assistance if a more detailed analysis is desired.

GLOSSARY

AC	-	Alternating Current
ACI	-	American Concrete Institute
ADT	-	Average Daily Traffic
AWS	-	American Welding Society
BART	-	Bay Area Rapid Transit
CIP	-	Cast-In-Place
CO_2	-	Carbon Dioxide
DC	-	Direct Current
FHWA	-	Federal Highway Administration
FTA	-	Federal Transit Administration
MTS	-	Maintenance Testing Specifications
NACE	-	National Association of Corrosion Engineers
NETA	-	InterNational Electrical Testing Association
NFPA	-	National Fire Protection Association
PVC	-	Polyvinyl Chloride
OSHA	-	Occupation Safety and Health Administration
SCADA	-	Supervisory Control And Data Acquisition
SEM	-	Sequential Excavation Method
TBM	-	Tunnel Boring Machine
USDOT	-	United States Department of Transportation

REFERENCES

American Concrete Institute (ACI) Committee 515, *A Guide to the Use of Waterproofing, Dampproofing, Protective, and Decorative Barrier Systems for Concrete, ACI 515.1R-79 (Reapproved 1985)*, American Concrete Institute, Farmington Hills, MI, 1986.

American Concrete Institute (ACI) Committee 224, *Causes, Evaluation and Repair of Cracks in Concrete Structures, ACI 224.1R-93 (Reapproved 1998)*, American Concrete Institute, Farmington Hills, MI, 1993.

American Railway Engineering and Maintenance-of-Way Association (AREMA), *Manual for Railway Engineering, Volume 1 and 4*, AREMA, 1999.

Arnoult, J. D., *Culvert Inspection Manual FHWA – IP – 86 – 2*, Federal Highway Administration, 1986.

Association Francaise des Travaux en Souterrains (AFTES) Working Group No. 14 (Maintenance and Repair of Tunnels), *Recommendations for the Treatment of Water Inflows and Outflows in Operated Underground Structures*, Tunnelling and Underground Space Technology, 4.3 (1989): 343-407. (Figures 4.7, 4.8, and 4.9 used with permission from Elsevier Science)

Barlo, T. J.; A. Zdunek, *Stray Current corrosion in electrified Rail Systems – Final Report, May 1995*, http://iti.acns.nwu.edu/projects/stray2 html, 12/7/2001.

Bickel, J.; E. King, and T. Kuesel, *Tunnel Engineering Handbook, Second Edition*, Chapman & Hall, New York, 1996.

Dell'Isola, A.J., *Value Engineering: Practical Applications...for Design, Construction, Maintenance & Operations*, R. S. Means Company, Inc., Kingston, MA, 1998.

Elliott, G. M.; M. R. Sandfort and J. C. May, *How to Prevent Tunnel Ice-up*, Tunnels & Tunnelling International, November 1995, pp. 42-43.

Emmons, P. H., *Concrete Repair and Maintenance Illustrated*, R. S. Means Company, Inc., Kingston, MA, 1994.

Engineering Services Division, *Life-Cycle Cost Analysis Instructional Manual*, Utah Department of Transportation, 1994

Haack, A.; J. Schreyer, and G. Jackel, *State-of-the-art of Non-destructive Testing Methods for Determining the State of a Tunnel Lining*, Tunnelling and Underground Space Technology, 10.4 (1995): 413-431.

Metro-North Commuter Railroad, *Manual for Maintenance and Inspection of Constant Tension catenary Systems*.

Narduzzo, L., *Tunnel Leak Remediation at the Toronto Subway*, Trends in Rock Mechanics, Geotechnical Special Publication No. 102, American Society of Civil Engineers Proceedings of Sessions of GeoDenver 2000.

National Fire Protection Association, *NFPA 502: Standard for Road Tunnels, Bridges, and Other Limited Access Highways*, 2001.

Newman, A., P.E., *Structural Renovation of Buildings, Methods, Details, and Design Examples*, McGraw-Hill, New York, NY, 2001.

Richards, J. A., *Inspection, Maintenance and Repair of Tunnels: International Lessons and Practice*, <u>Tunnelling and Underground Space Technology</u>, 13.4 (1998): 369-375.

Sperry Rail Service, *Rail Defect Manual*, Sperry Rail Service, 1968.

SYSTRA Consulting, *Electric Traction Catenary Inspection Field Manual*, AMTRAK, 1998.

U.S. Department of Transportation, Federal Railroad Administration – Office of Safety, *Code of Federal Regulations, Title 49, Track Safety Standards Part 213, Subpart A to F, Class of Track 1-5*, Simmons-Boardman Books, Inc., Omaha, NE, 2001.

U.S. Department of Transportation, Federal Railroad Administration – Office of Safety, *Code of Federal Regulations, Title 49, Part 220, Railroad Communications*, 2001.

U.S. Department of Transportation, Federal Railroad Administration – Office of Safety, *Code of Federal Regulations, Title 49, Part 236, Rules, Standards, and Instructions Governing the Installation, Inspection, Maintenance, and Repair of Signal and Train control systems, Devices, and Appliances*, 2001.

www.ingramcontent.com/pod-product-compliance
Lightning Source LLC
Chambersburg PA
CBHW080520110426
42742CB00017B/3187